# ISSUES IN SCIENCE EDUCATION

Jack Rhoton
Patricia Bowers
editors

National Science Teachers Association
National Science Education Leadership Association

Cover design by Barbara Richey

Copyright © 1996 by the National Science Teachers Association

Permission is granted in advance for reproduction of short portions of this book for the purpose of classroom or workshop instruction. To request permission for other uses, send specific requests to Special Publications, NSTA, 1840 Wilson Blvd., Arlington, VA 22201-3000.

NSTA Stock Number PB127X

Library of Congress Catalog Card Number: 95-73231

ISBN 0-87355-137-0

Printed in the U.S.A. by Bladen Lithographics

The National Science Teachers Association is an organization of science education professionals and has as its purpose the stimulation, improvement, and coordination of science teaching and learning.

# Table of Contents

## Science Education Reform

## Technology

## Science Education Research

## Assessment and Evaluation

## Science Education Leadership

## Effecting Change

## Professional Development

# Foreword

Science education reform requires leadership!

This publication appears at a time when reform in science education is as strong in some quarters as it has been in many, many years. At the national level, the *National Science Education Standards* have just been released. The *Benchmarks for Science Literacy* were published more than two years ago by the American Association for the Advancement of Science. There is a high level of support for science and math education from a number of federal agencies such as the National Science Foundation (NSF), Department of Education (DOE), National Aeronautics and Space Administration (NASA), and the Environmental Protection Administration (EPA). New and exciting curricula are still being developed with the support of NSF funds.

At the state and local levels, systemic initiative programs are in high gear. The Eisenhower Math and Science Education Act provides support for schools and districts with financial assistance for staff development. Over half of the states have completed new state frameworks or standards. Close working relationships and partnerships are being formed between schools and businesses, industries, and government agencies.

With so many programs and agencies involved, we sometimes forget that programs don't improve the quality of the education of our students—people do. And people require leadership and support. This sourcebook is about and for leaders, the people who make science education happen.

These are the leaders who will use the *Standards* and the *Benchmarks* in a wide variety of programs and settings to advance the quality of science education.

Leadership today in most progressive schools and organizations does not occur in the traditional top-down, authoritative structure many of us experienced in the past. It is "distributed" leadership that is found in many forms, locations, and levels performed by people in many different types of roles. The full-time, district level science supervisor is becoming an entity of the past. Instead, we are finding teachers who spend several hours a week in a leadership capacity in addition to their teaching role, staff from government agencies such as the Department of Energy who are providing leadership and support from outside the traditional school organization, university personnel who are working closely with school staffs, and scientists from industries and universities participating in school partnerships.

The concept of distributed leadership can be extended to everyone in the business of science education. If leadership is defined as an individual's ability and willingness to work with others to improve science teaching and learning, then virtually everyone in the science education community is included. We all become leaders, and we all accept responsibility for the reform of science education. We in NSELA salute you as a leader and welcome you to the resources in this volume and our Association.

*Harold Pratt*
NSELA Past-President, 1985–86

# About the Editors

**Jack Rhoton** is an associate professor of science education at East Tennessee State University, Johnson City, Tennessee. From 1992 to 1994, he chaired the study group committee for the Tennessee Science Curriculum Framework. Dr. Rhoton currently teaches science education at the undergraduate and graduate levels, and has also taught science at the elementary, middle, senior high school and college levels. He is the principal investigator for two major National Science Foundation (NSF) projects. He has received numerous awards for service and science teaching, and has been an active researcher in K–12 science, especially the restructuring of science inservice education as it relates to improved teaching practices. Dr. Rhoton is the editor of the *Science Educator*, a publication of the NSELA, and director of the Tennessee Junior Academy of Science. He also serves as editor of the TJAS *Handbook and Proceedings*.

**Patricia Bowers** is the associate director of the Center for Mathematics and Science Education at the University of North Carolina at Chapel Hill, where she teaches undergraduates and provides professional development training for math and science teachers. She also works closely with the UNC-CH Pre-College Program, which recruits under-represented groups into math and science fields. She has been the project director for numerous grants, including 12 Eisenhower grants, and has received awards for service and science education. Dr. Bowers was a Science, Mathematics, and Reading Coordinator at the system level and worked as a classroom teacher and guidance counselor at the school level. Dr. Bowers is currently president of the North Carolina Science Teachers Association, and secretary of the North Carolina Science Leadership Association. She serves on several committees for NSTA, and is a former district director and current board member of the NSELA.

# Preface

There is a growing consensus in our nation that education is in need of reform. Perhaps no area of the school curriculum has received more emphasis than the content of science education and how it gets presented to elementary and secondary students. This concern originates in more than 20 years of reports decrying the shortcomings of precollege science, as well as other aspects of the system believed to influence the teaching and learning of science.

As a consequence of these reports, the science education community has witnessed an unparalleled increase in education spending. States have legislated tougher standards, increased the number of science courses for high school graduation, made science courses more rigorous, increased the testing of both student and teacher, extended the school day and school year, and redefined the role of the school administrator. There is also a growing demand that schools examine their science programs in light of the *National Science Education Standards*.

This publication positively addresses issues and practical approaches needed to lay the foundation upon which science educators (at all levels) can work together to build effective science programs in our nation's schools. This book is an account of the issues that must be examined if science education reform is to move from rhetoric to reality. *Issues in Science Education* shares ideas, insights, and experiences of individuals ranging from science supervisors to university personnel to agencies representing science education. They discuss how to contribute to the success of school science and how to develop a culture that allows and encourages science leaders to continually improve their science programs.

The 29 chapters in *Issues in Science Education* are organized into seven sections. This organizational pattern is for the purpose of placing chapters within the context of a general theme. The intent of the book is not to provide an exhaustive coverage of each major theme, but rather to present a stimulating collection of chapters on relevant issues in science education. Numerous examples throughout the book illustrate the utility of topics to practioners as well as address general issues and perspectives related to science education reform.

Section I, "Science Education Reform," consists of six chapters that set the stage for the book by examining the issues associated with science education reform. Section II, "Technology," includes three chapters that illustrate how technology can be incorporated into the curriculum and how the use of technology can be used to promote student learning. Section III, "Science Education Research," contains three chapters that discuss the importance of basing curriculum and teaching decisions on research findings. Section IV, "Assessment and Evaluation," consists of four chapters that examine alternative methods of assessment and evaluation to deal with the enlarging variety of activities student engage in during their studies in science. Section V and VI, "Science Education Leadership" and "Effecting Change," made up of eight chapters, deal with the "Issues" that impact the day-to-day work of curriculum developers, instructional leaders, and science teachers. Finally, Section VII, "Professional Development," contains five chapters that address general issues and perspectives related to professional development.

Previous publications in this NSTA/NSELA series have been titled *Sourcebooks for Science Supervisors*. In 1994, however, recognizing the changing needs of science educators and the decline of the science supervisor, the National Science Supervisors Association changed its name to the National Science Education Leadership Association. In conjunction with the name change came a philosophical change as well. The organization changed its approach to be inclusive of not only traditional science supervisors, but also those who are assuming the duties of science education leader.

If we are to see paradigms of excellence in science education, all individuals responsible for quality science programs must use their knowledge, skills, and experiences to improve science teaching and learning. The task of reforming science education is simply too complex for any one person to tackle alone. Therefore, this work is directed at science teachers, science department chairs, principals, system wide science leaders, superintendents, university personnel, policy makers and other individuals who have a stake in science education. Moreover, administrators must create an atmosphere that supports and encourages participation in the improvement of science education. One of the greatest challenges of leadership is to develop a culture that creates "laboratories" that promote ongoing improvements. The final determinant of success in this effort will be measured through the quality of science programs delivered to our students.

Numerous examples throughout the book illustrate the utility of topics to practitioners and others interested in the issues of science education reform. Concepts like *Standards*, technology, action research in the classroom, assessment and evaluation, the science leader, multicultural science classrooms, constructivist leaders, laboratory safety, science teacher preparation, collaboration, elementary and secondary school science, and many other topics are placed in real world experience and combinations of original research.

This work would not have been possible without the help, advice, and support of a number of people. Most fundamentally, the members of the NSTA/NSELA Editorial Board—James Banks, France Brock, Karen Bullock, Michele Davis, Gerry Madrazo, Aaron Merik, LaMoine Motz, Susan Sprague, and Emma Walton—reviewed the manuscripts and made valuable suggestions for improvement. We could not have achieved our goal without their assistance, and we are grateful. Our appreciation is extended to Phyllis Marcuccio, Shirley Watt Ireton, Jennifer Hester, Catherine Lorrain-Hale, and Christina Frasch of NSTA Special Publications, for their invaluable help in the final design in which you are now reading. No volume is any better than the manuscripts that are contributed to it; we appreciate the time and efforts of those whose work lies within the covers of this book.

We also want to thank and acknowledge the support, help, and suggestions of the NSELA Board of Directors. Special thanks to Michael Jackson, past president, and Ken Roy, former Executive Director, for their suggestions and guidance in the early stages of this project. The support of Jane Hazen, president, and Patricia McWerthy, executive director, in the later stages of the project is gratefully acknowledged.

Finally, we would like to credit people who simply made room in their lives for us to do this work. We are indebted to the calm, good-natured support of the East Tennessee State University Division of Science Education office staff: Kelli Behuniak, Rachel Henry, and Janet Stanton. Each of these individuals did excellent work in word processing and typing the many drafts of each manuscript. And lastly, a special thanks to James Kevin, ETSU graduate student, for applying his expert editing skills to each manuscript.

*Jack Rhoton* and *Patricia Bowers*
Editors

# Introduction

Science leaders today come from many avenues: teacher, department head, supervisor, consultant, specialist, resource teacher, principal, curriculum director, superintendent, professor, scientist, researcher, and more. No longer can one person in an organization function with the sole responsibility of science education leadership. The role is too challenging, too complex, and too important. A team approach with several members who contribute to science education leadership is more appropriate for the realities of today's education workplace and society.

The NSELA/NSTA *Issues in Science Education* was created for a variety of individuals who function in some aspect of leadership in science education. With an explosion of current issues, such as the National Science Education Standards and other reform agendas, using technology to enhance learning, the role of research in improving instruction, improving assessment and evaluation in science education, professional development issues, and more, this book is stocked with valuable information and thought-provoking essays to arouse the interest of a wide audience. Articles written from a variety of perspectives—school district, university, and organization—will enable the Sourcebook to have a wide appeal.

The editors, Jack Rhoton and Pat Bowers, are to be congratulated for their superb efforts in this extremely time consuming task. They have edited an authoritative collection of articles and appropriately grouped them under common strands. This book is a "basic" for science leaders today. With today's vision of leadership involving shared power, share your power by sharing this resource with a colleague.

*Jane Hazen*
1995-96 NSELA President

# The Contemporary Reform of Science Education

Rodger W. Bybee

The science education community is now experiencing a transition in contemporary reform. In recent years, science educators have directed their attention to developing standards, whether at local, state, or national levels. But now the science education community has standards and benchmarks. Are standards and benchmarks enough? Can science educators declare the reform completed? Clearly, the answer to these questions is no. The even more difficult tasks of using standards and benchmarks to improve school programs and classroom practices are ahead. This position underlies the argument that the science education community is in a transition from a period when our attention has been intensely directed toward development of standards to a time when science educators have to clearly attend to the next phases of the reform.

This chapter addresses two aspects of the contemporary reform: (1) different dimensions of scientific literacy, and (2) different dimensions and difficulties of transitions in the contemporary reform. The need to understand scientific literacy in the context of school science programs complements the emerging need for curriculum, instruction, and assessments aligned with standards and benchmarks. Further, leaders should recognize the various issues attendant to the alignment and development of science education programs and practices with standards and benchmarks.

The chapter conveys the important and timely message that science education leadership will have to continue reform despite changing political support and mounting practical issues. Nobody said reform would be simple and easy. That is why science educators use the term *leadership* to describe the work of many individuals.

## UNDERSTANDING THE DIMENSIONS OF SCIENTIFIC LITERACY

Science educators have established consensus that the purpose of school science is to achieve scientific literacy. The idea of scientific literacy has been around for sometime. Table 1 presents examples of scientific literacy proposed by science educators over the years. In the contemporary reform, two documents clarify the view and specific details of scientific literacy: *Benchmarks for Science Literacy* (American Association for the Advancement of Science, 1993) and *National Science Education Standards* (National Research Council, 1996); see Table 2. These documents have consid-

Table 1. *Historical Examples of Categories Defining Scientific Literacy.*

| Science Literacy and the High School Curriculum[a] | What Is Unified Science Education? Program Objectives and Scientific Literacy[b] | Improving Indicators of the Quality of Science and Mathematics Education in Grades K-12[c] |
|---|---|---|
| Interrelationships between science and society | Nature of science | The nature of the scientific world view |
| Ethics of science | Concepts in science | The nature of the scientific enterprise |
| Nature of science | Processes of science | Scientific habits of mind |
| Conceptual knowledge | Values of science | Science and human affairs |
| Science and technology | Science and society | |
| Science in the humanities | Interest in science | |
| | Skills associated with science | |

*Note:* [a]The data in this column are from "Science Literacy and the High School Curriculum," by M. Pella, 1967, *School Science and Mathematics, 67,* pp. 346–356. Adapted with permission.

[b]The data in this column are from "What Is Unified Science Education?: Program objectives and scientific literacy," by Victor Showalter, 1974, *Prism II, 2*(2), pp. 1–6. Adapted with permission.

[c]The data in this column are from *Improving indicators of the quality of science and mathematics education in grades K–12,* by R. J. Murnane & S. A. Raizen, 1988. Washington, DC: National Academy Press. Adapted with permission.

Table 2. *Contemporary Examples of Categories Defining Scientific Literacy.*

| American Association for the Advancement of Science[a] | National Research Council[b] |
|---|---|
| The nature of science | Science as inquiry |
| The nature of mathematics | Physical science |
| The nature of technology | Life science |
| The physical setting | Earth and space sciences |
| The living environment | Science and technology |
| The human organism | Science in personal and social perspectives |
| Human society | History and nature of science |
| The designed world | Unifying concepts and processes |
| The mathematical world | |
| Historical perspectives | |
| Common themes | |
| Habits of mind | |

*Note:* [a]The data in this column are from *Benchmarks for Science Literacy,* by American Association for the Advancement of Science (AAAS), 1993, Washington, DC: Author. Adapted with permission.

[b]The data in this column are from *National Science Education Standards,* by National Research Council, 1996, Washington, DC: National Academy Press. Adapted with permission.

erable overlap and consistency for the content and they provide clear and elaborate definitions of scientific literacy (AAAS, 1995).

Science teachers, science educators, scientists, and engineers all contributed to the development of standards and benchmarks. Further, the processes of developing these documents involved critique and eventual consensus by significant numbers of individuals in the science education community. The science education leadership should use these documents' guidance in the next phases of contemporary reform. Science educators have to move beyond the analysis of these documents and use them to improve curriculum, instruction, and assessment and to inform professional development. It is inefficient for every state and local school district to completely reinvent standards and benchmarks. This position does not imply that states and school districts cannot adopt, adapt, and elaborate the national standards and benchmarks. The use of national standards and benchmarks at state and local levels, however, should include study and deliberation of the documents to clarify their intention and to avoid their inappropriate application to programs and practices.

Certainly, science educators who assume leadership roles have an obligation to understand the various domains of scientific literacy. Those domains include nature of science, physical, life, and earth sciences, nature of technology, science in society, and history of science; see Tables 1 and 2. The aforementioned standards and benchmarks provide conceptual frameworks and details of fundamental understandings for these various domains. Science educators also should understand various dimensions of scientific literacy as those dimensions specifically apply to curriculum, instruction, and assessment. The following sections elaborate dimensions of scientific literacy.

## Functional Scientific Literacy: The Vocabulary of Science

Scientific literacy includes vocabulary—the technical words associated with science and technology. Learners demonstrating functional literacy use scientific words appropriately and adequately. Relative to science and technology, learners should meet minimum standards of literacy as this term is usually defined. That is, given their age, stage of development, and educational level, learners should be able to read and write passages with scientific and technological vocabulary.

For years, school science programs, particularly science textbooks, have given this dimension of scientific literacy extraordinary emphasis and in many respects confused the achievement of scientific literacy with the accumulation of vocabulary and information—factoids about science and technology. Leaders in science education should recognize the error of overemphasizing vocabulary and strive for an appropriate balance on functional scientific literacy. The national standards and benchmarks have, for example, reduced but not eliminated technical words. These documents represent a first step toward an appropriate emphasis on scientific vocabulary in school programs.

## Conceptual and Procedural Scientific Literacy: Science As a Way of Knowing

In a 1995 article, Moore provided an insight for this dimension of scientific literacy:

> "Scientific literacy means relating those bits of information to the entire intellectual structure of the field. This is, facts are understood as parts of conceptua schemes" (p. 3).

In this quotation, Moore identified the conceptual dimension of scientific literacy. Although his discussion progressed to the importance of attitudes toward science, he anticipated a description of procedural scientific literacy and science inquiry by stating:

> "... But literacy in biology must be more than having a system for organizing the phenomena of living nature. It should imply also an active mind that seeks to use and relate all sorts of information" (p. 3).

In other works, Moore used "science as a way of knowing" to express both the conceptual and procedural dimensions of science (Moore, 1983, 1985a, 1985b, 1987).

Conceptual scientific literacy centers on the major ideas that form the disciplines of physical, life, and earth sciences. Table 3 displays an example of conceptual organizers from the national standards. This dimension of scientific literacy involves greater understanding of concepts that serve as the foundation for scientific disciplines.

Scientific inquiry exemplifies the procedural dimension of scientific literacy. The national standard on "Science as Inquiry" (NRC, 1995) highlights the abilities of in-

---

**Table 3. *An Example of Conceptual Scientific Literacy.***

**Life Science: Grades 9–12**

**Content Standard and Conceptual Organizers: The Cell**

As a result of their activities in grades 9–12, all students should develop an understanding of (a) the cell, (b) the molecular basis of heredity, (c) biological evolution, (d) the interdependence of organisms, (e) matter, energy, and organization of living systems, and (f) the nervous system and the behavior of organisms.

Fundamental concepts that underlie this standard include the following:

***Cells have particular structures that underlie their functions.*** Every cell is surrounded by a membrane that separates it from the outside world. Inside the cell is a concentrated mixture of thousands of different molecules which form a variety of specialized structures that carry out such cell functions as energy production, transport of molecules, waste disposal, synthesis of new molecules, and the storage of genetic material.

***Most cell functions involve chemical reactions.*** Food molecules taken into cells are broken down to provide the chemical constituents needed to synthesize other molecules. Both breakdown and synthesis are made possible by a large set of protein catalysts, called enzymes. The breakdown of some of the food molecules enables the cell to store energy in specific chemicals that are used to power the many functions of the cell.

***Cells store and use information to guide their functions.*** The genetic information stored in DNA is used to direct the synthesis of the thousands of proteins that each cell requires.

***Cell functions are regulated.*** Regulation of cells occurs both through changes in the activity of the functions performed by proteins and the selective expression of individual genes, allowing cells to respond to their environment and to control and coordinate the synthesis and breakdown of specific molecules, cell growth and division.

***Plant cells contain chloroplasts, the site of photosynthesis.*** Plants, and some other organisms, use solar energy to combine molecules of carbon dioxide and water into complex, energy-rich organic compounds. This process of photosynthesis provides a vital connection between the sun and the energy needs of living systems.

***Cells can differentiate, and complex organisms can develop from the generation of differentiated progeny of cell division.*** In the development of complex multicellular organisms, the progeny from a single cell form an embryo in which the cells differentiate to form the many specialized cells, tissues and organs that comprise the organism. This differentiation is controlled through the expression of different genes.

*Note:* From *National Science Education Standards,* by National Research Council (NRC), 1995, Washington, DC: National Academy Press. Adapted with permission.

---

quiry as well as understandings about inquiry; see Table 4. Science educators use terms from the 1960s and commonly express this aspect of scientific literacy as "the processes of science." In the contemporary reform of science education, the abilities of inquiry include, and have evolved beyond, the limited emphasis on processes, such as observation, inferences, hypothesis, and experiment. Table 4 indicates that the abilities of scientific inquiry includes

---

**Table 4.** *An Example of Procedural Scientific Literacy.*

**Science as Inquiry: Grades 9–12**

**Content Standard and Procedural Organizers:**

**Abilities Necessary to Do Scientific Inquiry**

As a result of their activities in grades 9–12, all students should develop the abilities of scientific inquiry and understandings about scientific inquiry.

Fundamental abilities and concepts that underlie this standard include:

*Identify questions and concepts that guide scientific investigations.* Students should formulate a testable hypothesis and demonstrate the logical connections between the scientific concepts guiding a hypothesis and the design of an experiment. They should demonstrate procedures, a knowledge base, and conceptual understanding of scientific investigations.

*Design and conduct scientific investigations.* Designing and conducting a scientific investigation requires introduction to conceptual areas of investigation, proper equipment, safety precautions, assistance with methodological problems, recommendations for use of technologies, clarification of ideas that guide the inquiry, and scientific knowledge obtained from sources other than the actual investigation. The investigation also may include such abilities as identification and clarification of the question, method, controls, and variables, the organization and display of data, the revision of methods and explanations, and the public presentation of the results and the critical response from peers. Regardless of the scientific investigations and procedures, students must use evidence, apply logic, and construct an argument for their proposed explanation.

*Use technology to improve investigations and communications.* Students' ability to use a variety of technologies, such as hand tools, measuring instruments, and calculators, should be an integral component of scientific investigations. The use of computers for the collection, analysis, and display of data also is a part of this standard.

*Formulate and revise scientific explanations and models using logic and evidence.* Student inquiries should culminate in formulating an explanation or model. In the process of answering the questions, the students should engage in discussions and arguments that result in the revision of their explanations. These discussions should be based on scientific knowledge, the use of logic, and evidence from their investigation.

*Recognize and analyze alternative explanations and models.* This standard emphasizes the critical abilities of analyzing an argument by reviewing current scientific understanding, weighing the evidence, and examining the logic, thus revealing which explanations and models are better and showing that although there may be several plausible explanations, they do not all have equal weight. Students should appeal to criteria for scientific explanations in order to determine which explanations are the best.

*Communicate and defend a scientific argument.* Students in school science programs should develop the abilities associated with accurate and effective communication including writing and following procedures, expressing concepts, reviewing information, summarizing data, using language appropriately, developing diagrams and charts, explaining statistical analysis, speaking clearly and logically, constructing a reasoned argument, and responding to critical comments through the use of current data, past scientific knowledge, and present reasoning.

*Note:* From *National Science Education Standards*, by National Research Council (NRC), 1995, Washington, DC: National Academy Press. Adapted with permission.

---

the aforementioned processes of science and give greater emphasis to cognitive abilities, such as using logic, evidence, and extant knowledge, to construct explanations of natural phenomena.

## Multidimensional Scientific Literacy: The Contexts of Science

Scientific literacy extends beyond vocabulary, conceptual schemes, and procedural methods of the physical, life, and earth sciences. The educational goal of achieving scientific literacy includes other understandings about science. Science educators have to help learners develop perspectives of the science disciplines that include, for example, the history of scientific ideas, the nature of science, and the role of science in society.

One aspect of multidimensional scientific literacy centers on disciplines and includes the history and nature of science and technology. For example, students should develop some understanding of the history of biology, chemistry, physics, and the earth sciences. In addition, students should learn about the nature of science and technology. The national standards and benchmarks included the history and nature of science and technology.

Another aspect of multidimensional scientific literacy refers to understandings about science in personal and social life or, to use a contemporary theme, science, technology, society (S-T-S). Science exists in society and certainly has connections to many contemporary issues. An essential part of achieving scientific literacy involves helping learners understand the limits and possibilities of science in their personal lives and in society.

The view of scientific literacy described here incorporates dimensions that have had various proponents and emphases. But we have historically overstated the case for one dimension, such as the structure-of-the-dis-

cipline approach in the 1960s or the S-T-S theme in the 1980s, or we have overemphasized one aspect of scientific literacy, such as scientific vocabulary or process skills. The contemporary vision of scientific literacy should represent an appropriate balance among functional, conceptual, procedural, and multidimensional aspects of this powerful idea.

## Nominal Scientific Literacy: Science from Students' and Teachers' Perspectives

Underlying each dimension of scientific literacy is what I term *nominal scientific literacy*. Learners do not demonstrate complete and accurate expressions of the qualities that define the aforementioned dimensions of scientific literacy. Nominal scientific literacy means that learners associate terms, ideas, and issues with the general area of science and technology, but these associations represent misconceptions, naive theories, or incomplete understandings. A common definition of the term *nominal* suggests that the relationship between the learners' understandings and abilities, such as those described in national standards and benchmarks, is small and insignificant. There are, at best, only token understandings and abilities, and they bear little relationship to the goal of scientific literacy. Science educators, such as supervisors, teacher educators, curriculum developers, assessment specialists, and science teachers, have the task of designing programs and implementing practices to help learners move beyond their nominal levels and achieve higher levels of scientific literacy in all the aforementioned dimensions—functional, conceptual, procedural, and multidimensional.

Table 5 summarizes the characteristics of scientific literacy discussed in this section. Achieving higher levels of scientific literacy means that students will develop appropriate knowledge, skills, abilities, and understandings associated with the different dimensions of scientific literacy described in this section. Such a view is quite different from an overemphasis on one dimension or a single sentence that many seek when they ask for a definition of scientific literacy. The standards and benchmarks provide full and detailed definitions of scientific literacy. If only learners demonstrated the understandings and abilities in these documents, science educators would have made consider-

---

**Table 5.** *Dimensions of scientific literacy.*

### Nominal Scientific Literacy

Identifies terms, questions, as scientific but demonstrates incorrect topics, issues, information, knowledge, or understanding.

Misconceptions of scientific concepts and processes.

Inadequate and inappropriate explanations of scientific phenomena.

Current expressions of science are naive.

### Functional Scientific Literacy

Use scientific vocabulary.

Define scientific terms correctly.

Memorize technical words.

### Conceptual and Procedural Scientific Literacy

Understands conceptual schemes of science.

Understands procedural knowledge and skills of science.

Understands relationships among the parts of a science discipline and the conceptual structure of the discipline.

Understands organizing principles and processes of science.

### Multidimensional Scientific Literacy

Understands the unique qualities of science.

Differentiates science from other disciplines.

Knows the history and nature of science disciplines.

Understands science in a social context.

---

able progress in science education and as a society.

## UNDERSTANDING THE DIMENSIONS OF CONTEMPORARY REFORM

If achieving higher levels of scientific literacy is the goal and leaders in the science education community understand the various aspects of scientific literacy, then it seems important to have a map of the reform territory. Such a map for science educators will help identify their location, means of movement, and the direction and difficulties of travel. In the science education community, the variety of reports, standards, and studies on the contemporary reform of science education presents a confusing array of maps, destinations, and directions discussed at national, state, and local levels. A simple framework is quite helpful for locating and clarifying different efforts in the geography of contemporary reform. The framework uses the terms *purpose*, *policy*, *programs*, and *practice* to identify various dimensions of reform.

### The Purpose of Science Education: Achieving Scientific Literacy

Science educators have expressed many aims, goals, and objectives in various documents, such as national standards, state frameworks, school syllabi, and teaching lessons (DeBoer, 1991; Bybee, 1993; Bybee & DeBoer, 1994; DeBoer & Bybee, 1995). For this discussion, the term *purpose* refers to universal goal statements of what science education should achieve. Such statements are abstract and apply to all concerned with science education. Achieving scientific literacy is a statement of the purposes of science education. The strength of this purpose statement lies in its widespread acceptance and agreement among science educators. The weakness is found in its ambiguity concerning specific situations in science education. For example, what does the goal of achieving scientific literacy mean for a third grade teacher? a high school chemistry teacher? a teacher educator? a policy maker? a curriculum developer? The answer, of course, varies for different situations—hence the need for more concrete statements of scientific literacy and the adaptation of those statements for various factions of the science education community. Statements that are based on the general purpose, but which address more specific situations, introduce the role of policies.

## Policies for Science Education: National, State, and Local Standards

Policy statements are concrete translations of the purpose for various components of the science education community. Documents that give direction and guidance, but are not actual programs, serve as policies. Examples of policy documents include course plans for high school earth science, district syllabi for K–12 science, state frameworks, and national standards and benchmarks. Likewise, college or university requirements for undergraduate teacher education and state and national frameworks for assessing scientific literacy also fall into the category of policies for science education. At the national level, examples of policy documents include the *National Science Education Standards* (NRC, 1996) and *Benchmarks for Science Literacy* (AAAS, 1993). The contemporary reform is poised at the critical point between completing standards and benchmarks and implementing programs that align with those documents.

## Programs for Science Education: Curriculum, Assessment, and Professional Development

Programs include the actual curriculum materials, textbooks, and courseware based on the policies. Programs are unique to grade levels, disciplines, and aspects of science education, such as teacher education or a middle school science program.

School science programs may be developed by national organizations, such as Bio-logical Sciences Curriculum Study (BSCS), and marketed commercially, or they may be developed by states or school districts. Who develops the materials is not the defining characteristic; the fact that schools, colleges, state agencies, and national organizations have programs aligned with policies such as the national standards is the important feature of this aspect of reform.

## Practices of Science Education: Teaching and Learning Science

Practice here refers to the specific actions and processes of teaching science in schools, colleges, or universities. The practices of science education include the personal interactions between teachers and students and among students, as well as the roles and uses of assessment, educational technologies, laboratories, and myriad other methods of teaching science. In the framework described here, implementing new classroom practices implies they would be consistent with policies, and programs would be designed to achieve a purpose, such as scientific literacy.

Improving the practices of teaching science centers on the most individual, unique, and fundamental aspect of the framework. Science educators can propose new goals, design new standards, syllabi, and scope and sequence charts, and they can develop new curriculum materials, but the critical aspect of the contemporary reform is improving teaching and enhancing learning in science classrooms.

## Understanding Leadership Roles and the Current Situation in Reform

The various leadership roles in the science education community can be identified by locating the work of individuals in the framework. Obviously, leaders in national, state, and local positions will vary in their current tasks. However, the tasks and direction of change will involve aspects of the aforementioned framework. The *Na-*

*tional Science Education Standards* may serve as a general example of where science education is in the contemporary reform and the difficulties of scaling-up to the next phase of the contemporary reform.

## LOOKING BEYOND STANDARDS

The *National Science Education Standards* (NRC, 1996) will provide a powerful set of policies to guide the improvement of science education programs and classroom practices. As important and challenging as the development of the *Standards* has been (and state and local standards for that matter), they represent only one step in the progress of standards-based systemic reform of science education. Although many different steps may follow publication and dissemination of the standards, a vision beyond standards seems particularly important—provision of science *programs* and *practices* aligned with *National Science Education Standards*.

## Transforming Standards to Programs and Practices

Will widespread dissemination of national standards provide adequate guidance for the implied transformation of curriculum materials, instructional practices, and assessment strategies? They may. However, they may not if science educators and science teachers fail to understand the assumptions upon which the standards were developed (LeMahieu & Foss, 1994) and consequently assume that dissemination of the standards alone will result in the changes they propose. By design, the standards do not provide complete programs, practices, and assessments. Practitioners with technical expertise to read, interpret, and implement all dimensions of the standards will have to implement the policies as school science programs and innovative instructional practices.

Common sense and educational research support the conclusion that the science education community needs more than stan-

dards to initiate and sustain the kind of changes outlined in the *Standards*. There are several reasons for this conclusion. First, the standards document presents practitioners with a formidable amount of information. Although the *Standards* are well thought out, provide clear and accurate descriptions of content, teaching, assessment, program, and system, and include some examples, they do not lend themselves to easy interpretation.

Second, for the most part, the standards do not provide clear and compelling descriptions of processes for translating the standards. This observation should not come as a surprise because describing the process of implementing the standards was not the purpose of the document. Implementing the standards will require concerted efforts by the leadership in science education.

Third, the changes implied by the national standards present a complex array of interdependent factors involving content, teaching, assessment, professional development, school science programs, and systemic reform. Individuals and groups commonly review standards based on the single factor most closely aligned with that individual's or group's professional interest, and based on the review, infer what the standards may mean. Although examining and understanding the *Standards* as an integrated set of policies is a professional obligation, this position is neither intuitively obvious nor commonly practiced. It certainly should be, though, for leaders in the contemporary reform.

Finally, practitioners are bound by their current views of science education programs and practices. These views are often contrary to the spirit of the standards and benchmarks and also sustained by many commercial publishers, school boards, administrators, and communities. In the constructivist sense, those views have to be challenged and shown to be inadequate to the current situation, and new ideas—programs and practices—that are meaningful, feasible, and usable have to be proposed.

The aforementioned factors imply the need for several things: (1) discussions and dialogue about standards, (2) illustrations of what the standards mean in terms of school science programs (including assessment), and (3) what standards mean for teaching practices, professional development, and systemic support for science education.

One of the most important needs that teachers in science education can fulfill centers on discussions and dialogue about standards. Adequate translation of standards to school science programs and classroom practices requires that individuals understand standards as an integrated set of policies. Further, there are subtitles of emphasis, depth, and domains and dimensions of scientific literacy that require study and deliberation of the standards.

There is also a need to provide expressions of the standards through exemplars that bring the standards to life so they can guide those responsible for transforming and implementing standards. From the perspective of school science programs, the national standards were not framed to present a coherent curriculum, coordinated instructional sequences, and associated assessments. The content standards, rather, present outcomes—not the means to achieve the outcomes. There is a need for examples of both the process and products of transforming the national standards and benchmarks into programs and practices.

## UNDERSTANDING THE DIMENSIONS AND DIFFICULTIES OF REFORM

Assuming the above section made the case for moving beyond the standards, science educators will recognize the difficulties in scaling-up these efforts in the contemporary reform. Here the intention is to encourage support of the standards and reform through understanding of the roles science educators play and the issues they must resolve. In general, we have to look at the research and insights of individuals, such as Fullan and Stiegelbauer (1991) and Hall (1989). The leadership in science education will find a recent book by Anderson and Pratt (1995) particularly helpful.

In addition to these references and resources, the following descriptions of various aspects of reform may help science educators to identify the various dimensions and difficulties of reform. The assessments represent an armchair analysis, not a scientific study, but considerable support and evidence confirms the framework.

## Dimensions of Contemporary Reform

Table 6 presents various dimensions of reform. The left column summarizes the perspectives of purpose, policy, program, and practice. The top row identifies six dimensions of reform: time, scale, space, duration, materials, and agreement. Reviewing the table provides a general sense of both the dimensions and difficulties of the reform effort. Science educators can ask questions such as how long does it take to form policies, such as national standards or state frameworks? Once a new program is implemented, how long does it continue in a school system? What is the location of particular efforts such as curriculum reform or policy formation? Although individuals can find exceptions to the framework, it does present a general picture of various dimensions of reform efforts.

Table 6 gives an overall picture of the reform effort and identifies our present situation in the contemporary reform. Science educators should get an operational sense of what scaling-up means as they progress from a period of policy development to the improvement of programs and implementa-

**Table 6. Dimensions of Contemporary Reform.**

| Perspectives | Time (For actual change to occur.) | Scale (Number of individuals involved.) | Space (Scope and location of the change activity.) | Duration (Once change has occurred.) | Materials (Actual products of the activity.) | Agreement (Difficulty reaching agreement among participants.) |
|---|---|---|---|---|---|---|
| **Purpose**<br>• Reforming goals.<br>• Establishing priorities for goals.<br>• Providing justification for goals. | **1-2 Years** To publish document. | **Hundreds** Philosophers and educators who write about aims and goals of education. | **National/Global** Publications and reports are disseminated widely. | **Year** New problems emerge and new goals and priorities are proposed. | **Articles/Reports** Relatively short publications, reports, and articles. | **Easy** Small number of reviewers and referees. |
| **Policy**<br>• Establishing design criteria for programs.<br>• Identifying criteria for instruction.<br>• Developing frameworks for curriculum and instruction. | **3-4 Years** To develop frameworks, legislation, reviewers. | **Thousands** Policy analysts, legislators, supervisors, and areas. | **National/State** Policies focus on specific areas. | **Several Years** Once in place, policies are not easily changed. | **Book/Monograph** Longer statements revisions. | **Difficult** Political negotiations, trade-offs, and revisions. |
| **Program**<br>• Developing materials or adopting a program.<br>• Implement the program. | **3-6 Years** To develop a complete educational program. | **Tens of Thousands** Developers, field-test teachers, students, textbook publishers, software the program. | **Local School** Adoption committees. | **Decades** Once developed or adopted, programs last for extended periods. | **Books/Courseware** Usually several books for students and teachers. | **Very Difficult** Many factions, barriers, and requirements. |
| **Practices**<br>• Changing teaching strategies.<br>• Adapting materials to unique needs of schools and students. | **7-10 Years** To complete implementation and staff development. | **Millions** School personnel, public. | **Classrooms** Individual teachers. | **Several Decades** Individual teaching practices often last a professional lifetime. | **Complete System** Books plus materials, equipment and support. | **Extraordinarily Difficult** Unique needs, practices, and beliefs of individuals, schools, and communities. |

Table 7. *Difficulties of Contemporary Reform.*

| Perspectives | Risk to Individual School Personnel | Cost in Financial Terms to School | Constraints Against Reform for School | Responsibility for Reform by School Personnel | Benefits to School Personnel and Students |
|---|---|---|---|---|---|
| **Purpose**<br>• Reforming goals.<br>• Establishing priorities for goals. | Minimal | Minimal | Minimal | Minimal | Minimal |
| **Policy**<br>• Establishing design criteria.<br>• Identifying criteria for instruction.<br>• Developing frameworks for curriculum and instruction. | Moderate | Moderate | Moderate | Moderate | Moderate |
| **Program**<br>• Developing materials or adopting a program.<br>• Implement the program. | High | High | High | High | High |
| **Practices**<br>• Changing teaching strategies.<br>• Adapting materials to unique needs of schools and students. | Extremely High | Extremely High | Extremely High | Extremely High | Extremely High |

tion of practices. The next phases of contemporary reform will take longer, involve more individuals, move closer to schools and classrooms, last longer, involve more materials and equipment, and present more difficulties when it comes to reaching agreement and actually implementing improvements designed to help students achieve scientific literacy.

## Difficulties of Contemporary Reform

Table 7 describes other aspects of contemporary reform. Again, the left column includes the perspectives of purpose, policy, program, and practice. This table uses issues of risk, cost, constraints, responsibilities, and benefits and considers these difficulties in terms of school districts, school personnel, and students—the place and people ultimately responsible for reforming programs and practices. The analysis presented in Table 7 indicates that, although essential, purpose statements and policy documents have minimal and moderate influence on reform. The science education community is now approaching the phases of reform where risk, cost, constraints, personal responsibilities, and benefits, are all high or extremely high. Clearly, this means the leadership in science education has significant challenges.

## CONCLUSION

This chapter introduces and clarifies the unifying purpose of the reform, that is, achieving scientific literacy. The discussion of scientific literacy focused on the dimensions of functional, conceptual and procedural, and multidimensional.

The current phase of reform indicates that leaders in science education must shift from the development of standards defining scientific literacy to the development of programs and practices aligned with standards and thus designed to achieve scientific literacy. Although logical and seemingly simple, the direction of this transition will require more time, individuals, materials, and agreement than development of standards and benchmarks. In addition, the cost, risks, constraints, and responsibilities, all move to higher levels. In the end, the benefits to students and society also are extremely high. Although the current situation presents significant challenges, we have the experience, understanding, and leadership to meet those challenges.

## References

American Association for the Advancement of Science. (1993). *Benchmarks for science literacy.* Washington, DC: National Academy Press.

Anderson, R., & Pratt, H.A. (1995). *Local leadership for science education reform.* Dubuque, IA: Kendall/Hunt.

Bybee, R.W. (1993). *Reforming science education: Social perspectives and personal reflections.* New York: Teachers College Press.

Bybee, R.W., & DeBoer, G.E. (1994). "Research as goals for the science curriculum." In *Handbook of research in science teaching and learning,* by D. Gabel (Ed.). New York: Macmillan

DeBoer, G.E. (1991). *A history of ideas in science education.* New York: Teachers College Press.

DeBoer, G.E., & Bybee, R.W. (1995). The goals of science curriculum. In *Redesigning the science curriculum: A report on the implications of standards and benchmarks for science education,* by R.W. Bybee & J.D. McInerney (Eds.), pp. 71–74. Colorado Springs, CO: Biological Sciences Curriculum Study (BSCS).

Fullan, M., & Stiegelbauer, S. (1991). *The new meaning of educational change* (2nd ed.). New York: Teachers College Press.

Hall, G.E. (1989). "Changing practice in high school: A process, not an event." In *High school biology: Today and tomorrow,* by W.G. Rosen (Ed.). Washington, DC: National Academy Press.

LeMahieu, P.G., & Foss, H.K. (1994). Standards at the base: What are the implications for policy and practice? *The school administrator: journal of the american association of school administrators, 51*(5), 16–22.

Moore, J.A. (1983). *Science as a way of knowing. Evolutionary biology* (Vol. 1). Baltimore, MD: American Society of Zoologists.

Moore, J.A. (1985a). *Science as a way of knowing. Human ecology* (Vol. 2). Baltimore, MD: American Society of Zoologists.

Moore, J.A. (1985b). *Science as a way of knowing. Genetics* (Vol. 3). Baltimore, MD: American Society of Zoologists.

Moore, J.A. (1987). *Science as a way of knowing. Developmental biology* (Vol. 4). Baltimore, MD: American Society of Zoologists.

Moore, J.A. (1995, January). Cultural and scientific literacy. *Molecular biology of the cell, 6*, 1–6. Baltimore, MD: American Society of Zoologists.

Murnane, R.J., & Raizen, S.A. (1988). (Eds.) *Improving indicators of the quality of science and mathematics education in grades K–12.* Washington, DC: National Academy Press.

National Research Council (NRC). (1995). *National science education standards.* Washington, DC: Author.

Pella, M. (1967). Science literacy and the high school curriculum. *School science and mathematics, 67*, 346–356.

Showalter, V. (1974). What is unified science education?: Program objectives and scientific literacy. *Prism II, 2*(2), 1–6.

## Author Note

**Rodger W. Bybee** is Executive Director, Center for Science, Mathematics, and Technology Education of the National Research Council. He was the Chair, Content Working Group, the NSES&A project of the NRC. He is a former associate director of the Biological Sciences Curriculum Study (BSCS) and Chair, Content Working Group, the National Science Education Standards and Assessment Project of the National Research Council. He has designed science curriculum for elementary, middle, high school, and college levels. He has written articles and books on science teaching and curriculum reform.

# Role of the School Principal In Science Education Reform

J. Preston Prather

Science is now widely accepted as a basic curricular component, and there is increasing public pressure for schools to provide all students with an adequate science education. This has prompted a variety of efforts for improvement of science teaching, including the employment of school science specialists, provision of continuing education for teachers, implementation of lead teacher programs, and development of a variety of curricular schemes. This chapter examines an additional essential consideration, the role of the school principal in science education reform.

## OVERVIEW

Several studies of the roles of principals in school improvement were examined with attention to implications for science education, and a golden thread of consensus ran through much of the literature: The ideal role for school principals is leadership through partnerships with teachers in shared vision-making, decision-making, and responsibility for improving the schools.

Several studies in science education supported that ideal. For example, Mechling and Oliver (1982) concluded that involvement of the school principal is essential to

the success of any elementary science program. Prather, Hartshorn, and McCreight (1988) described an extension of the ideal to include joint preparation of teachers and their principals and curriculum supervisors through instruction in science and science teaching:

> By working hand in hand with their teachers, the administrators gained an understanding of how science is taught and the sort of support their teachers need to do an effective job. Likewise, the teachers gained invaluable insight into the administrative resources and limitations relevant to their work. The resultant clear channels of communication alone should help those school systems overcome many hurdles in their quest for better science education ( p. 460).

Reporting on the outcomes of two state-wide programs that incorporated the joint preparation of teachers and their principals as a team, Rhoton, Field, and Prather (1992) concluded that "the key to success was the central involvement of the school principal as a coequal with teachers in ... inservice education" (p. 14). To test the concept, three teams were formed without principals, and each of the rest consisted of a principal and teachers from the same school. "Among the three teams that substituted another teacher for the administrator, inservice activities and other functions were substantially more limited than with the other teams" (Prather, 1993). When teachers and principals joined forces to plan and implement improved science programs, however, "they were adequate to the task," Rhoton et al. (1982, p. 22) reported, and "many teachers perceived the principals' participation as an expression of appreciation for the importance of their teaching roles" (p. 19). Perhaps a comment by a former teacher and principal, and later co-director of the Principals' Center at Harvard University, best reflects the basis for that success: "When a principal shares a foxhole with a teacher, collegiality, staff development, and higher morale are possible outcomes. As much can be gained from stumbling together as from succeeding" (Barth, 1988, p. 641).

As Prather et al. (1988) implied, an ideal way for principals to share the foxhole with science teachers is to involve themselves as coequals in preparation for and implementation of reform efforts. This could be accomplished by enrolling with teachers in summer science institutes and academic-year programs, joining in the planning and implementation of reforms, occasionally team-teaching a science lesson with a teacher, and sharing in the conduct of inservice programs. The resulting interaction with teachers would provide shared insight into the unique needs of science teaching. It would also prepare principals for participatory leadership in the improvement of science teaching, which is their appropriate and essential role in educational reform.[1] "Although a principal may not have specific knowledge of every curricular area taught in the school," Smith and Andrews (1989) concurred, "his [her] knowledge should at least embrace the general trends in each subject area" (p. 14). This may not be easy

for some principals to do, however, because of varying and sometimes conflicting perceptions of their role in school operations.

## PERCEPTIONS OF THE PRINCIPAL'S ROLE

"The school principalship has been the subject of hundreds of studies over the past 30 years," Smith and Andrews (1989) noted, and "the central role of the principal has been viewed, variously, as building manager, administrator, politician, change agent, boundary spanner, and instructional leader" (p. 1). This appears to be the case for elementary and secondary school principals alike. Furthermore, the review of literature revealed little if any difference in general perceptions of the teacher-principal relationships at various grade levels. Consequently, no differentiation is made between elementary and middle school principals in this analysis of the principal's role in science education reform.

In earlier times, the principal played a dual role as a teaching principal, or part-time teacher and part-time administrator. Typically, the principal was a long time co-worker of the teaching staff. This close relationship resulted in most or all teachers being closely involved in local decision making. The dual roles were compatible, and teachers generally perceived the principal as an advocate for them and their programs whenever school matters were taken to the school board or superintendent's office. But for many principals, Lietner (1994) explained, the perception has changed over time to one of *competing* expectations:

From the beginning of the modern, non-teaching principalship, the principal was saddled with competing roles of principal as administrator and as instructional manager or leader ... an increasing portion of the principal's time is devoted to administrative rather than instructional activities.

Nevertheless, principals…want to be involved in instructional activities because they believe that the instructional role, not the administrative role, influences student learning. (p. 219)

As increasing administrative duties moved principals out of the classroom, an uneasy tension ensued. Teachers wondered about their principal's priorities. Was this new breed of principal their colleague and advocate in determining and achieving school goals, or was the priority now the imposition of externally imposed standards and practices? The schism widened as principals were put into, or perceived as being put into, an adversarial role in supervision and teacher evaluation that "provides a means for direct control over teachers" (Leitner 1994, p. 223).

More recently, the emergence of career-ladder plans, teacher centers, and the recommendation of the Carnegie Task Force on Teaching as a Profession for "lead teachers" suggests there should be a diminution of the principal's role as instructional leader. Discussions of these efforts usually reflect a political orientation and a concern for authority and power. (Smith & Andrews, 1989, p. 1)

Clearly, principals are caught in a dilemma. On the one hand, they feel pressure to give additional priority to administration. On the other hand, they feel a need to give more attention to instructional leadership. In addition, their job is affected by some educational policymakers who tend to view concepts of business management as transferable to school administration. In the stable technical environment of businesses and industries, a manager is expected to establish a division of labor, control resources and employee behavior for maximum efficiency, protect productive personnel from external disturbances, and otherwise control conditions to assure ongoing productivity.

When transferred to the educational arena, the business management model places principals in the role of direct management of teacher behavior to assure efficient production of educated children. This outlook contributes to a perception of teachers' roles in learning as that of technicians responsible for carrying out externally prescribed practices rather than that of professionals responsible for diagnosing and fulfilling students' learning needs. This was reflected in the development of externally planned science programs and science activities kits in the 1960s and 70s that were designed to be implemented in the classroom by teachers in much the same way that a manufacturing procedure would be carried out on an assembly line by a factory worker.[2] However, Leitner (1994) noted, a school's technical environment is much less stable than that of a business. It is difficult for principals to maintain narrowly focused school goals because of unpredictable factors that may affect instructional outcomes under differing conditions. Consequently:

Principals avoid direct manipulation of teacher behavior and instruction to increase efficiency since it is difficult to know or to verify which changes would increase efficiency. (p. 223).

An efficient, businesslike manner is considered appropriate, it seems, when it comes to shaping impersonal raw materials into products or providing a detached service, but a much more versatile and interpersonal approach is required when it comes to educating children. This is especially the case among parents (Lindle, 1989). Principals are directly responsible for the quality of school instruction and thereby play a pivotal role in the type of learning environment that a school provides for children. Whatever the expectations of their position when it comes to matters of classroom teaching, most principals are far more comfortable in a collaborative role of instructional leadership rather than in one of direct control.

## TEACHER-PRINCIPAL RELATIONSHIPS

A review of recent research on the roles of principals indicated a similar ideal among teachers and principals alike: Leadership, rather than management or structural control, was the preferred role for a principal within the school. Many principals attempt to avoid the risks of more direct forms of control over teachers and teaching, which may lead to teacher dissatisfaction (Hoy, Tarter, & Forsyth, 1978) and a sense of professional alienation (Cox & Wood, 1980). Rather, they prefer to use cultural and interpersonal mechanisms to gain teacher consensus on school goals and coordinate teachers' instructional practices (Leitner, 1994).

The trend, it seems, is swinging back toward a teaching principalship, or a collegial relationship of principals and teachers in the setting and implementation of school goals. This was reflected in a statewide survey of teachers that revealed a clear preference of principals as leaders rather than managers:

> "Teachers do have clear expectations of the role they want a principal to play in creating a positive learning environment for students. ... Providing effective instructional leadership ... is clearly the role that teachers desire for principals" (Williams, 1988, p. 113).

Those principals who are most successful in providing instructional leadership, Hord (1988) concluded, "use a model of shared leadership and a collaborative approach where teachers have important roles" (p. 11). Barth (1988) concurred that teachers should share in school leadership and suggested ways that principals may encourage them, ranging from "articulating the goal" (p. 640) to "sharing responsibility for failure" (p. 641) to

"admitting ignorance" (p. 642). The latter suggestion, he noted, would imply a need for help that teachers may provide, and this could be a disarming incentive for collaboration. In a study involving 190 schools in Hong Kong, Cheng (1994, p. 300) adapted a leadership forces model developed by Sergiovanni (1984) to classify strengths of school principals according to the following five categories of leadership:

1. *Human leadership:* high sensitivity to needs of others, supportive.

2. *Structural leadership:* clear planning, holding people accountable for results.

3. *Political leadership:* capable of handling and resolving conflicts.

4. *Educational leadership:* encourages professional development for teachers.

5. *Symbolic leadership:* symbolizes school values, celebrates school successes to build morale.

In a similar study of the principal's role in instructional leadership in 27 schools in the Northwestern United States, Leitner (1994) used an operational definition of principal involvement adapted from a study by Hallinger and Murphy (1985). Leitner (1994, p. 221) grouped behaviors of effective principals into three categories:

1. defining the school's mission

2. managing the instructional program

3. promoting a positive school climate

In a study involving several hundred principals and more than 2,500 teachers, Smith and Andrews (1989, pp. 9-19) classified a strong principal as one with proficiency as:

1. a resource provider

2. an instructional resource

3. a communicator

4. a visible presence

Cheng (1994) concluded that "the stronger the leadership, the more the schools are perceived effective in terms of productivity, adaptivity, and flexibility" (p. 312). School authority was perceived as less hierarchical with strong leadership, he found, and teachers working with strong principals were more satisfied and reported greater opportunity to participate in determination of school policies and allocation of resources. Leitner (1994) concurred that "school improvement is a complex process that involves a host of factors which principals can influence, but not control" (p. 236). He concluded that principals who support and foster teacher participation are consistently more successful in school improvement. Instructional reform is critically dependent upon shared leadership by principals and teachers (Barth, 1988; Cheng, 1994; Hord, 1988; Krug, 1992; Leitner, 1994; Mechling & Oliver, 1982, 1983, 1988; Shoemaker & Fraser, 1981), and review of literature on the role of teachers and principals in school improvement clearly revealed that both groups prefer a sharing relationship.

## IMPLICATIONS FOR SCIENCE EDUCATION REFORM

Principals can make or break a science program. They are in a position to promote or denigrate change in curriculum and instruction, and their support or lack of support is a critical element in a school science program's success or failure. Unfortunately, Mechling and Oliver (1982) observed, "principals appear to have been neglected in our national efforts to improve science education, and our programs have suffered because of this neglect" (p. xi). Some efforts have been made to involve principals in reform efforts, but few have been systematic.

"Few administrator training programs … go beyond the one-shot approach," Hord (1988) declared. "Much staff development for principals has been characterized as a jumble of quick-fix sessions that are topic-specific, short-term, content-loaded, and held out of the district" ( p. 11). Clearly, "an administrator trying to be an instructional leader has had little direction in determining just what it means to do so" (Hallinger & Murphy, 1985, p. 217).

More long term, systematic involvement with teachers and teaching is needed to acquaint principals with the problems and needs of the various subjects included in the curriculum. As a study of the principal's role in elementary science education indicated, principals not familiar with the unique requirements of science teaching can neither relate effectively to the instructional needs of their science teachers nor confidently promote the improvement of science education:

During the heyday of science education in the 1960s and 1970s, … few [principals] participated in federally funded efforts to improve science education. Instead, teachers were the targets. Typically, elementary teachers went off to summer and academic-year programs to learn about new science curricula and teaching techniques. Many became strong advocates for the improvement of science education, only to find, on returning to their home schools, that the coach was bent on playing the same old game in the same old way. The spirit of science, the new ways of teaching, and the enthusiasm that these programs inspired hadn't reached the principals—and they were calling the plays. … Even the most enthusiastic teachers began to lose heart. They needed administrative support, and they weren't getting it. Science education was in limbo. (Mechling & Oliver, 1983, p. 14)

Only actual involvement in the science teaching arena will provide the types of experiences needed by principals who were prepared in non-science fields. Otherwise, science education is destined to remain in limbo. Surely all principals, like the teachers they work with, want the best for their students. However, they cannot be expected to know how to provide the best without an operational understanding of problems related to teaching a subject. Whereas some topics may be taught in a traditional lecture-textbook-demonstration-recitation manner, science cannot. (It is debatable how effective this approach may be in any subject.) Yet a great majority of science classes are taught in a traditional, textbook-centered, student-passive manner with lecture being the dominant mode (Weiss, 1987). Lacking the understanding and support—or prodding—of their principal, many teachers have either not obtained or not maintained the resources, skills, and continuing education needed for reform of science and mathematics[3] teaching.

Science teaching requires an extensive variety of supplies and materials for use in developing hands-on science activities to illuminate the concepts to be taught in science. Adequate planning time and space for procuring materials, planning and setting up manipulatives, and testing activities is essential to assure the safety and instructional efficacy of lesson plans. Also, some activities may involve materials, such as life forms or perishable substances, that require special attention, and long-term experiments such as those designed to study the life cycle may require that instructional apparatus be maintained in place for weeks or months. A working acquaintance with these needs will enable principals to plan more effectively for instructional improvement.

Hands-on science teaching methods have been rated superior for promotion of outcomes such as science content mastery, criti-cal thinking ability, and problem solving skills (Shymansky, Kyle, & Alport, 1983). Development of hands-on skills, whether related to teaching or driving or flying or mountain climbing or hundreds of other things, can be attained only through physical learning experiences. Sooner or later, an aspiring pilot, for example, must take the airplane controls in hand—under the watchful eyes of an experienced instructor, of course—and gain flying skills through hours of experience in flight. Given sufficient interaction of a learner and an experienced pilot in actual flight, a successful solo flight may be expected; otherwise, the learner is likely to meet with disaster. No airport manager would risk the disaster of allowing a person to go on a solo flight if that person had never actually flown a plane, no matter how many books or videotapes or flight simulators the person had worked with. Only supervised experience in flight can equip a person with the skills and confidence to deal with the quick decision-making that may be required in the air. Similarly, no principal should expect a teacher to launch science teaching reforms without adequate inservice education and support. The risk is too great. Teachers who are expected to teach hands-on science lessons, for instance, without a chance to work with people experienced in that methodology face a no less serious likelihood of failure in successfully attaining and implementing the required teaching skills. The potential cost in teacher disillusionment and the loss of learning opportunities for the school's students is obvious.

Successful reform of science education will require the central involvement and leadership of principals in the process of researching, planning, and implementing instructional improvements. Collaboration with teachers to include shared leadership in school goal setting and allocation of resources for instruction and professional development has shown the greatest potential for success, and this must be a top priority for schools dedicated to reform in science education.

## CONCLUSION

Clearly, the ideal role for school principals in the reform of science education is collaborative leadership with teachers in shared vision-building, decision-making, and responsibility for improvement of school science teaching. For most principals, the immediate need is to get involved in their school science programs in order to gain an understanding of the problems of science teaching and learn how they may support reforms. Mechling and Oliver (1982), whose work constitutes a comprehensive reference for administrators interested in science education reform, recommended that a principal fulfill the following roles:

1. Science Leader: "Principals are urged to ... demonstrate positive attitudes toward science, and to communicate their interest in science to teachers and others" (p. xi).

2. Curriculum Analyst: "Principals need to find out about their own science programs—the goals, the teaching strategies used by teachers, typical learning experiences, evaluation procedures, and so on" (p. xii).

3. A Force In the Selection or Development of a New Science Curriculum: "Principals should ... be the driving force, the catalyst, to get the process going and keep it on track" (p. xii).

4. Provider of the Wherewithal: "An important role of the principal is to seek funds to purchase science supplies and equipment and cover costs of staff development" (p. xiii).

5. Provider of Inservice Instruction: "Principals must take the lead in providing inservice experiences in science for their teachers ... by ... participating, actively, in inservice programs themselves" (p. xiii-xiv).

6. Monitor of Progress in Science Programs: "Principals must give their science programs periodic checkups to determine their vitality and continuing effectiveness" (p. xiv).

7. Troubleshooter: "Principals must be able to deal effectively with problems that arise in the science curriculum—problems such as teachers' slouching off their science teaching responsibility" (p. xiv).

Fulfillment of those roles will require that principles join with their teachers in teacher enhancement programs and other activities necessary to gain an operational understanding of the nature and needs of science teaching. Principals with little prior science preparation may require formal science instruction beyond that available through inservice programs. This will require a major effort of many principals, but it is the key to successful reform of science teaching. "There's only one way to improve your science class," Mechling and Oliver (1988) concluded, "and that's to improve your principal's science quotient" (p. 13). That extra effort will be a small price to pay for the opportunity to achieve the goal of improved scientific literacy.

### Footnotes

[1] Many authors cited in this chapter wrote from the perspective of general instructional improvement rather than reform of science teaching in particular, but the consensus was essentially the same for any subject area: the role of the principal in school reform is shared leadership.

[2] Such programs were sometimes referred to, often derisively, as teacher-proof curricula. Clearly, such an attitude was not conducive to positive self-perceptions among teachers.

[3] Most of the science education issues discussed in this chapter are equally applicable to the mathematics education arena. There is increasing awareness of the benefits of in-

terdisciplinary instruction in science and mathematics, and helping to implement this change in the school curriculum is one of the most important tasks that principals face.

## References

Barth, R.S. (1988). Principals, teachers, and school leadership. *Phi Delta Kappan*, 69, 639–642.

Cheng, Y.C. (1994). Principal's leadership as a critical factor for school performance: Evidence from multi-levels of primary schools. *School Effectiveness and School Improvement*, 5, 299–317.

Cox, H., & Wood, J. (1980). Organizational structure and professional alienation: The case of public school teachers. *Peabody Journal of Education*, 9, 1–6.

Hallinger, P., & Murphy, J. (1985). Assessing the instructional management behavior of principals. *Elementary School Journal*, 86, 217–247.

Hord, S.M. (1988). The principal as teacher. *Journal of Teacher Education*, 39(3), 8–12.

Hoy, W., Tarter, C., & Forsyth, P. (1978). Administrative behavior and subordinate loyalty: An empirical assessment. *Journal of Educational Administration*, 16, 29–38.

Krug, S.E. (1992). Instructional leadership: A constructivist perspective. *Educational Administration Quarterly*, 28, 430–433.

Leitner, D. (1994) Do principals affect student outcomes: An organizational perspective. *School Effectiveness and School Improvement*, 5, 219–238.

Lindle, J.C. (1989). What do parents want from principals and teachers? *Educational Leadership*, 47(2), 12–14.

Mechling, K.R., & Oliver, D.L. (1982). *The principal's role in elementary school science*. Washington DC: National Science Teachers Association.

Mechling, K.R., & Oliver, D.L. (1983). The principal's project: Promoting science among elementary school principals. *Science and Children*, 21(3), 14–16.

Mechling, K.R., & Oliver, D.L. (1988). Your basic principal of science. *Science and Children*, 25(6), 12–14.

Prather, J.P. (1993). A model for inservice science teacher enhancement through collaboration of rural elementary schools and universities. In *Excellence in educating teachers of science*, P. Rubba, L. Campbell, & T. Dana, (Eds.), pp 131–149. Columbus, OH: Ohio State University.

Prather, J.P., Hartshorn, R., & McCreight, D. (1988). A team leadership development program: The Elementary Science Education Institute. *Education*, 108, 454–461.

Rhoton, J., Field, M., & Prather, J.P. (1992). An alternative to the elementary school science specialist. *Journal of Elementary Science Education*, 4(1), 14–25.

Sergiovanni, T.J. (1984). Leadership and excellence in schooling. *Educational leadership*, 41(4), 4–13.

Shoemaker, J., & Fraser, H. (1981). What principals can do: Some implications from studies in effective teaching. *Phi Delta Kappan*, 63, 178–182.

Shymansky, J., Kyle, W., & Alport, J. (1983). The effects of the new science curricula on student performance. *Journal of Research in Science Teaching*, 20, 397–404.

Smith, W.F., & Andrews, R.L. (1989). *Instructional leadership: How principals make a difference*. Alexandria, VA: Association for Supervision and Curriculum Development.

Weiss, I. (1987). *Report of the 1985-86 national survey of science and mathematics education*. Research Triangle Park, NC: Research Triangle Institute.

Williams, J.C. (1988). How do teachers view the principal's role? *NASSP Bulletin*, 72, 111–113.

## Author Note

**J. Preston Prather** is an assistant professor of science education at the University of Virginia. He has received several academic awards, including the NTSA's History of Science Education Award and the Outstanding Educator of the Year award presented by the American Association of Educators of Teachers in Science (AETS) in 1995.

# Science Teacher Preparation As A Part Of Systemic Reform In The United States

Robert E. Yager

Failures of most school programs to result in the graduation of students ready and able to assume citizenship responsibilities in a scientific and technological world represents a continuing problem in science education. Cognitive science studies are cited to illustrate that many of the best students who continue as physics and engineering majors at colleges are unable to do anything except repeat concepts and work problems where equations and directions are provided for solutions. Features necessary to gain real engagement of student minds and real learning are identified. All of this information about past failures is used to suggest needed changes in science teacher education programs. If changes are to be made as pathways to produce better learners, it is obvious that teachers need a different preparatory program—one that results in students with the ability to use concepts and processes on their own and in new situations. A basic problem is that most newly prepared science teachers merely mimic their own experiences as students, which suggests that knowledge and skills are transmitted directly from teacher to student. Some needed changes in typical preservice programs are identified and discussed.

## THE HISTORY OF SCIENCE EDUCATION REFORM MOVEMENTS

Perhaps the single most distinguishing feature of science education reform for the 1990s is the use of "systemic" as an adjective to reform. Some have insisted that restructuring is a more accurate reflection of the changes needed rather than reform. Others have defined the systemic idea to represent such a significant change in terms of the reform that we can claim a shift in paradigms (Cossman, 1967; Kuhn, 1970; Roy, 1983).

Hurd (1985) has pointed out that there have been dozens of national reform movements in science education in our history as a nation. All have been attempts to make the content more applied and related to the lives of students except for the reforms following Sputnik. The reforms from 1957–70 all followed the position articulated by Zacharias (the architect of the Physical Science Study Committee [PSSC] course), which proclaimed that science will be interesting and appropriate for all if it is portrayed in a way known to scientists. Bruner (1960) simultaneously posited that anything can be taught in some intellectually honest form to any learner regardless of the stage of intellectual development. These positions set the national reform effort in a new direction that also attracted over two billion dollars of financial support from the government. By the end of the 1970s, disillusionment with this approach certainly appeared among the general citizenry as well as among most practicing teachers, if not among the science education leadership that had been a part of the efforts. However, some advances were made, especially in the development of new course materials and in terms of new models for enhancing teaching skills.

Early in the 1980s it was apparent that economic competitiveness was a more serious blow to our national prestige and the possible longevity of our international leadership than Sputnik provided 25–30 years earlier. The new calls included science for all (suggesting that the focus on pure science of the 1960s was not), less is more (emphasizing that covering too many skills and concepts in a course distracted from learning), joining science and technology (reinforcing the perception that technology was less important than a focus on the natural world thereby reducing the historic separation of technology for average and below average students and science for the college preparatory student), and dissolution of the "layer-cake" curriculum where year-long courses in biology, chemistry, and physics had become the common course sequence in U.S. high schools and the high school program had been moved to the junior high school, often with year-long courses in life, physical, and earth science. The focus of most curriculum efforts of the 1990s has been to integrate the sciences around common themes and ideas downplaying discipline traditions and focusing on real-world problems.

The early 1980s brought a new interest in, and support for, research into human learning. Cognitive science has been created as a discipline in the United States with more research funds earmarked for such studies

than for any other field of education. Cognitive scientists of today were formerly psychologists. National Science Foundation (NSF) funding in the United States is more readily attained when the effort is cast with the rubric cognitive science. Such a focus on science was also judged to be superior to supporting research in science education per se. Education too often is seen as contributing to the problems of science in K–12 settings.

Initially the cognitive science researchers were intent upon unlocking the workings of the human brain. They were anxious to know how the best brains worked, to permit educators to deal more effectively with those brains that were not so effective. Actually these "non-working minds" were possessed by most high school graduates. Extensive studies by Miller and his colleagues at the University of Northern Illinois revealed that 90 percent were scientifically illiterate (Miller, Suchner, & Voelker, 1980). Recent studies following the extensive reform efforts during the past decade indicate that the percentage of science illiterates has grown (Miller, 1989; National Assessment of Educational Progress, 1989).

Cognitive scientists selected university physics majors for initial study, presumably because they were thought to possess the best minds. They were the best high school graduates in terms of completing all high school science courses with distinction (upper 10 percent). But, they also enjoyed their study and were successful with even more advanced courses in college. If we could understand how the brains of such people operated, surely we would begin to have an awareness about the workings of the human mind that could help with less successful minds.

After careful experimentation with physics majors (and later engineering students), it was a great surprise to find that the minds of these most successful college students, who were studying science and engineering to pursue careers in these fields, were really no different from the minds of elementary and secondary students who could not operate well in the worlds of science and engineering. Fully 85 percent of the physics majors (Champagne & Klopfer, 1984; Mestre & Lochhead, 1990) and 90 percent of the engineering students could not use the information and skills they seemed to possess in any real-world, problem-solving situation. Admittedly, the tests the cognitive scientists devised were difficult; they were meant to be. They hoped to analyze the processes that the best students used to solve the problems or to explain the phenomena. Obviously it was a shock to find that the best students of secondary schools and successful physics and engineering college students were unable to use the concepts and processes in new situations in the real world away from classrooms and laboratories.

This led the cognitive scientists to study physics and engineering students further to explain their misconceptions, ideas, and inability to interpret and use the materials and skills—even though they could recite and perform in classrooms and laboratories (Helm & Novak, 1983; Novak, 1987). In every case studied, the explanations and interpretations offered arose from some personal construction—some interpretation of natural phenomena that these most successful students had constructed themselves. Personal experience was more powerful in operating in the real world than were the ideas and skills students had learned in formal classrooms and laboratories.

This research also caused many to look anew at the basic features of science. Simpson wrote extensively about the meaning of science, especially a meaning with relevance to teaching. His definition says, "science is an exploration of the material universe in order to seek orderly explanations (generalizable knowledge) of the objects and events encountered: *but these explanations must be testable*" (Simpson, 1963,

p. 81). Science is first of all an *exploration* of the universe. It requires a personal curiosity—a wonderment for the world around us. Basically, all people are curious, some more than others. Unfortunately, the typical school seems to discourage such exploration. Hence, the school program discourages the first feature of real science.

The second ingredient of science is the attempt to explain the events and objects encountered during exploration. The formation or creation of such explanations is essential for science to occur. It is a phenomenon that is discouraged in most school curricula, especially in traditional science programs. Again it seems that common practice is alien to the basic nature of science.

According to Simpson, the third and final feature of science is the testing of the explanations personally formulated. To be scientific, explanations must be *testable*. It is possible to offer all kinds of explanations: creative ones, ones dependent upon the supernatural, ones that defy experimentation. Such explanations may be appropriate for other human enterprises such as creative writing, art, or religion. However, if an explanation of events and/or objects arising from explorations of the universe cannot be tested for its validity, there can be no science experience (Pittendrigh & Tiffany, 1957).

The 1980s also saw interest in Science-Technology-Society (STS) as a major reform initiative. STS represents a shift in paradigm and affects goals, instruction, curriculum, and assessment—all critical incidents in the process of education. STS was supported by Norris Harms' landmark work called *Project Synthesis* (1977). This research centered upon the justification (and goals) for K–12 science. These included

1. *Science for meeting personal needs*. Science education should prepare individuals to use science for improving their own lives and

for coping with an increasingly technological world.

2. *Science for resolving current societal issues.* Science education should produce informed citizens prepared to deal responsibly with science-related societal issues.

3. *Science for assisting with career choices.* Science education should give all students an awareness of the nature and scope of a wide variety of science and technology-related careers open to students of varying aptitudes and interests.

4. *Science for preparing for further study.* Science education should allow students who are likely to pursue science academically as well as professionally to acquire the academic knowledge appropriate for their needs.

Harms ended his report with the following analysis and concerns:

> The goals of preparing the majority of students to use science in their everyday lives, to participate intelligently in group decisions regarding critical science-related societal issues, and to make informed decisions about potential careers in science and technology are equally as important as the goal of preparing a minority of students for more advanced coursework in science. Thus, a new challenge for science education emerges. The question is this: "Can we shift our goals, programs, and practices from the current overwhelming emphasis on academic preparation for science careers for a few students to an emphasis on preparing all students to grapple successfully with science and technology in their own everyday lives as well as to participate knowledgeably in the important science-related decisions our country will have to make in the future?" (Harms, 1981, p. 119).

## Constructivism and Instruction as Pivotal to Current Reform

The cognitive science research has resulted in a new look at constructivism as the most appropriate learning theory—to connect the problems arising from the reforms of the 1960s and emphasizing the importance of instruction over curriculum *per se*. Most of the past 40 national efforts in reform have stopped with guidelines for reforming the curriculum or the actual new course outlines and textbooks that illustrated the reform movement of the 1960s. Of course, these were complete with inservice programs for teachers, but research shows such efforts to be flawed. The teachers did not learn, probably for some of the same reasons that students fail to learn in K–12 classrooms. Teachers and students were (are) too often taught to remember, repeat, recall, and regurgitate information and skills—just because they are "important and fundamental." No real context is provided where these important intended outcomes can be seen and experienced as useful and personally meaningful.

The intellectual engagement of all students is basic to successful teaching and a precursor to learning. Students do not learn by reading information from textbooks or listening to teacher lectures where success is measured by the degree such information is remembered. Perrone (1994) has identified eight ways in which student minds become engaged:

1. Students help define the content.

2. Students have time to wonder and to find a particular direction that interests them.

3. Topics have a "strange" quality—something common seen in a new way, evoking a "lingering question."

4. Teachers permit—even encourage—different forms of expression and respect students' views.

5. Teachers are passionate about their work. The richest activities are those "invented" by teachers and their students.

6. Students create original and public products; they gain some form of "expertness."

7. Students *do* something—e.g., participate in a political action, write a letter to the editor, work with the homeless.

8. Students sense that the results of their work are not predetermined or fully predictable.

Once intellectual engagement is realized, learning is likely to occur. Since real learning rarely occurs as a result of typical instruction occurring in most K–12 and college science classrooms, new approaches are needed (Lochhead & Yager, 1996; Mestre & Lochhead, 1990). Typical teaching places premiums on students who can repeat, recall, and regurgitate science concepts and process skills that usually provide the curriculum frameworks. As late as 1990, Mestre and Lochhead proclaimed that the goals for science education can be defined in terms of two dimensions, namely concepts and processes (Mestre & Lochhead, 1990). Perhaps both are unimportant dimensions if a real-world context is not provided.

Reinsmith (1993) has delved more deeply into real learning—learning that goes beyond repeating, recalling, and regurgitating. He has described learning in the following ways:

1. Learning takes place as a result of a process much like *osmosis*.

2. Authentic learning comes through trial and error.

3. Students learn only what they have some proclivity for or interest in.

4. No one formally learns something unless he/she *believes* he/she can learn it.

5. Learning cannot take place outside an appropriate context.

6. Real learning connotes *use*.

7. No one knows how a learner moves from imitation to intrinsic ownership, i.e., from external modeling to internalization and competence.

8. The more learning is like play, the more absorbing it will be.

9. For authentic learning to happen, time should occasionally be wasted, tangents pursued, side-shoots followed up.

10. Traditional tests are very poor indicators of whether an individual has really learned something.

## Science Teaching Can Be a Science

Teachers intimately involved with current reforms must be reflective in thinking about their teaching. They must question their behaviors and actions and hypothesize about how they might impact learning both negatively and positively. This suggests science teaching can be a science as defined by Simpson. Teachers must raise questions for which possible answers can provide the basis for observations, data collection, and data analysis to determine the validity of the idea that was proposed.

Teachers involved with current reforms must exemplify constructivist teaching practices (Yager, 1991):

1. Seeking out and using student questions and ideas to guide lessons and whole instructional units.

2. Accepting and encouraging student initiation of ideas.

3. Promoting student leadership, collaboration, location of information, and taking actions as a result of the learning process.

4. Using student thinking, experiences, and interests to drive lessons (even if this means altering teacher's plans).

5. Encouraging the use of alternative sources for information both from written materials and experts.

6. Using open-ended questions and encouraging students to elaborate on their questions and responses.

7. Encouraging students to suggest causes for events and situations and encouraging them to predict consequences.

8. Encouraging students to test their own ideas by answering their own questions, guessing as to causes, and predicting certain consequences.

9. Seeking out student ideas before presenting teacher ideas or before studying ideas from textbooks or other sources.

10. Encouraging students to challenge one another's conceptualizations and ideas.

11. Using cooperative learning strategies that emphasize collaboration, respect individuality, and use division of labor tactics.

12. Allowing adequate time for reflection and analysis.

13. Respecting and using all ideas that students generate.

14. Encouraging self-analysis, collection of real evidence to support ideas, and reformulation of ideas in light of new experiences and evidence.

## Current Teacher Preparation and the Need for College Major Reforms

All the changes in science education (the national movement in K–12 schools) suggest the imperative of major shifts in programs designed to prepare new teachers as well as programs designed to assist inservice teachers to initiate and expand changes currently advocated. Too often these changes are more difficult to make than those in schools. Project 2061 was the first and largest of the reforms of the 1990s, and it was predicated on the notion that major changes (shifts in paradigm) can not be made quickly. In fact the project is conceived as a 75-year effort; 2061 is the year that Halley's Comet will next be visible on earth. However, project leaders indicate that some results and indications may be apparent after a decade of thinking, trials, and re-trials.

The Carnegie Foundation funded the American Association for the Advancement Of Science (AAAS) with a major grant to study science in the liberal arts in colleges and universities (AAAS, 1990). Many see *the* major problem slowing reforms in K–12 science education to be college science courses and teaching. College science professors fail to understand and/or to use the research arising from cognitive science studies of the last decade. The AAAS report calls for reform of college science teaching, which will also provide a major advantage for preparing new science teachers:

> The goals of science in liberal education cannot be accomplished in a single course or in a few, isolated science courses. What is needed is a coherent integrated program of liberal studies, designed and implemented by the entire institution. The science courses themselves will require reconceptualization, as will the current structure of the curriculum. The traditional survey courses and concern about "coverage"

have no place in the new curriculum described here. Rather, it is critical that liberal education in the sciences become a well-integrated part of a broader liberal education program.

Faculty from the humanities, fine and practical arts, and social sciences are essential collaborators with scientists in framing the definition of "scientific literacy" and designing curricula to provide it. Nothing will happen unless their goodwill and commitment are forthcoming. Natural science faculties must take the initiative by assembling relevant data, generating dialogue, and fostering collaboration among faculty throughout an institution. In addressing issues of technology in society, the natural science faculty will need to work with faculty in the social sciences, humanities, and with engineering faculties on campuses that have them (AAAS, 1990, pp. xvii-xviii).

Science teacher education has consisted of an undergraduate degree with three basic ingredients, namely a major in science (or, at some institutions, *one* of the sciences), the general education requirements of the college, and a professional education component. The science major is usually set by faculty in the various areas of science and usually has not differed significantly from the various programs leading to graduate work in one of the sciences and/or preparation for medicine or other related health fields. Often this major (with supporting courses in mathematics and other related science fields) consists of work approaching 60 semester hours—two full years of preparation. Typical programs in no way meet the challenges and the recommendations of the AAAS report on needed changes in collegiate science programs. Often, the general education requirement (communication skills, physical education, social studies, literature, foreign language, and humanities)

includes 30 to 45 hours of additional credit. Professional education courses often include 18 to 30 hours of credit. When a student completes the general educational and professional sequence in education, there is usually little or no opportunity for electives. In fact, this situation has led many to opt for masters of teaching (MAT) degrees to permit preservice teachers to complete a few graduate courses in science, to spread out the professional courses over more semesters, and to complete some electives and/or special support areas.

Yager (1986) has analyzed this problem as options are proposed to prepare STS teachers and others ready to initiate the current reforms in schools. Basically the problem of science teacher education has never been conceived as one that science educators can solve alone. In a real sense systemic thinking, planning, and action are needed. Certainly new approaches are needed in defining a science major (as proposed in the AAAS report), reconceptualizing the general education program, and creating a more extensive program in science education. There is little hope for major changes unless we abandon some "sacred cows" including completion of typical science majors, the general education requirement, which assumes no preparation coming from middle or secondary schools, and the notion that teachers should all experience some history/philosophy of education, education psychology, instructional design, human relations, special education, and evaluation/measurement. To assume that one science educator on a given campus can teach one methods course and supervise teaching practica and student teaching experiences is ludicrous.

Yager's analysis speaks to the problem of thinking about courses as the only organizers for teacher education programs. Haberman (1984) has recently provided an interesting history for the course concept in higher education. He raises serious questions of organizing a meaningful program with such a design. It could be in the best interests of science education if experimental colleges developed teacher education "freer" of courses. For most it may be appropriate to think of facets of a complete program in terms of individual courses or a series of courses. These courses would, of necessity, be quite different from standard science courses found in most departments and from the teacher education sequence described earlier. More of the program could resemble general education experiences—at least those in the most innovative institutions.

Yager (1986) has developed a rationale for the inclusion of program features for a science teacher ready for the current reforms. The program allows for greater specialization and for the inclusion of electives and some collegiate general education offerings. The rationale also assumes that the program leading to licensure as a science teacher is but the beginning of continuous education leading to refinement as a practicing teacher. A 70-hour program was offered as the minimum as movement to science teacher education reform is undertaken. This minimal program is offered in five components for a bachelors' degree and provisional licensure:

| | |
|---|---|
| Basic Science, including some work in all traditional disciplines | 30 hours |
| Applied Science, including engineering and other vocationally oriented fields | 10 hours |
| Social Studies, including sociology, geography, and economics | 10 hours |
| History, Philosophy, and Sociology of Science | 10 hours |
| Courses focusing on the use of science/technology concepts and skills to solve problems (i.e., project approach) | 10 hours |
| **Total** | 70 hours |

It is apparent that such a "major" surpasses the total hours required for the typical discipline-based science major that most preservice science teachers commonly complete at the current time. However, it is believed that some of the courses in the recommended 70 hour major would also fulfill many of the usual courses in general education. Only the skill areas and perhaps literature/humanities/art would be additional requirements (if they were not already satisfied as part of a typical high school program).

The teacher education sequence remains an important part of a total program for preparing teachers for the reforms of the 1990s. Certainly some of the recommendations and positions mentioned initially should be included and, where possible, in a science education context. When a proper and meaningful context is not provided, many ideas and concepts remain abstract and unused. The research affecting teacher education in general could fall in that realm if the persons involved with special programs do not help with the context and build upon it throughout the entire program.

The best teachers ready for leadership in reforming schools can more readily be prepared when appropriate staff (i.e., science educators) are involved throughout a four- or five-year program. Certainly it is rare to find a student who can select a variety of courses to meet requirements and who sees or senses their relationships or value. A sensitive faculty member can help with the synthesis, the connections, and the insights.

Early and frequent experiences in schools with pupils is a vital part of any meaningful and successful preservice teacher education program. Work in schools should be a part of the real society, it should provide a means of dealing with youth and an opportunity to motivate students to take action, to use knowledge, to debate, and to make informed decisions. A preservice teacher education program must aim toward science teachers who are able and comfortable with teaching approaches required for current reforms.

## CONCLUSION

The current reform movements in schools are upon us. As we all strive to prepare more and better teachers to face tasks only dreamed about five years ago, it is apparent that science teacher education must be improved and strengthened as soon as possible. Surely traditional programs headed by generalists and persons unaware of current research and new problems have no chance to resolve these issues. Unfortunately, such programs may be the cause of the failure of science education. As more and more teachers are needed, often to fill positions defined in a most traditional way, the situation will get worse. If new teachers cannot bring new ideas, we all lose. If new teachers are more equipped for schools, students, and programs of the 1960s, we seem hopelessly lost. Major changes are essential. Educating a better total citizenry for the 21st century may provide our only real hope for survival of the human race. Can we do less than try with all our might and intellect?

Obviously the job of inservice programs is the same as that facing science educators charged with preparing new teachers. This is where systemic approaches to the problem are again in evidence. It is not enough to focus on new course structures, a given group of inservice teachers, a program for preparing teachers for licensure, or the involvement solely of educators in the process. Systemic change, to be lasting and de-serving of the claim of "systemic," must include all stakeholders: community leaders, administrators, teacher educators, business/industry leaders, parents, representatives from professional societies, accountability and licensure officials, and many more. Yes, the task is awesome, but it is one we must all undertake. The quality of our future and indirectly our survival may depend on the results of our combined wisdom and actions.

## References

American Association for the Advancement Of Science (AAAS). (1990). *The liberal art of science: Agenda for action (The report of the project on liberal education and the sciences)*. Washington, DC: Author.

Bruner, J. (1960). *The process of education*. New York: Vintage Books.

Champagne, A.B., & Klopfer, L.E. (1984). Research in science education: "The cognitive psychology perspective." In *Research within reach: Science education,* by D. Holdzkom & P.B. Lutz (Eds.), pp. 171–189. Charleston, WV: Research and Development Interpretation Service, Appalachia Educational Laboratory.

Cossman, G.W. (1967). *The effects of a course in science and culture designed for secondary school students.* Unpublished doctoral dissertation, University of Iowa, Iowa City.

Haberman, M. (1984). The origin of the university course. *Journal of Teacher Education, 35*(4), 52–54.

Harms, N.C. (1977). *Project Synthesis: An interpretative consolidation of research identifying needs in natural science education. (A proposal prepared for the National Science Foundation.)* Boulder, CO: University of Colorado.

Harms, N.C. (1981). Project Synthesis: Summary and implications for action. In *What research says to the science teacher, Volume 3,* by N.C. Harms & R.E. Yager (Eds.), pp. 113–127. Washington, DC: National Science Teachers Association.

Helm, H., & Novak, J.D. (Eds.). (1983). *Proceedings of the international seminar on misconceptions in science and mathematics*. Ithaca, NY: Cornell University, Department of Education.

Hurd, P. DeH. (1985, February). *Update on science education research: The reform movement*. Paper presented at a meeting of The Appalachia Educational Laboratory, Inc., RDIS Training Workshop, San Francisco, CA. (Ed 260 942)

Kuhn, T. (1970). *The structure of scientific revolutions* (2nd ed.). Chicago: Chicago University Press.

Lochhead, J., & Yager, R.E. (1996). Is science sinking in a sea of knowledge? A theory of conceptual drift. In *Science/technology/society as reform in science education*, by R.E. Yager (Ed.), 25–38. Albany, NY: State University of New York Press.

Mestre, J.P., & Lochhead, J. (1990). *Academic preparation in science: Teaching for transition from high school to college*. New York: College Entrance Examination Board.

Miller, J.D. (1989, April). *Scientific literacy*. Paper presented at the meeting of the American Association for the Advancement of Science, San Francisco, CA.

Miller, J.D., Suchner, R.W., & Voelker, A.M. (1980). *Citizenship in an age of science: Changing attitudes among young adults*. New York: Permagon Press.

National Assessment of Educational Progress (NAEP). (1989). *Science objectives, 1990 assessment*. Booklet No. 21-S-10. The Nation's Report Card. Princeton, NJ: Educational Testing Service.

Novak, J.D. (Ed.). (1987). *Proceedings of the second international seminar on misconceptions in science and mathematics*. Ithaca, NY: Cornell University, Department of Education.

Perrone, V. (1994). How to engage students in learning. *Educational Leadership, 51*(5), 11–13.

Physical Science Study Committee (1960). *Physics*. Boston, MA: Heath.

Pittendrigh, C.S., & Tiffany, L.H. (1957). *Life: An introduction to biology*. New York: Harcourt Brace Jovanovich.

Reinsmith, W.A. (1993). Ten fundamental truths about learning. *The National Teaching & Learning Forum, 2*(4), 7–8.

Roy, R. (1983, May 19). Math and science education: Glue not included. *The Christian Science Monitor*, pp. 36-37.

Simpson, G.G. (1963). Biology and the nature of science. *Science, 139*(3550), 81–88.

Yager, R.E. (1986). Restructuring science teachers education programs as they move toward an S/T/S focus. In *1985 AETS Yearbook: Science, technology and society: Resources for science educators*, by R.K. James (Ed.), pp. 46–55. Columbus, OH: Association for the Education of Teachers in Science and Science, Mathematics, and Environmental Information Analysis Center.

Yager, R.E. (1991). The constructivist learning model: Towards real reform in science education. *The Science Teacher, 58*(6), 52–57.

## Author Note

**Robert E. Yager** is a professor of science education at the University of Iowa. His work has resulted in nearly 500 publications; he has advised over 100 Ph.D. students and has headed seven national organizations in science education, including NSTA. He has received many honors, including NSTA's Distinguished Service to Science Education Award and the Carleton Award.

# Beyond Infomercials and Make-and-Take Workshops:
# Creating Environments for Change

Susan L. Westbrook
Laura N. Rogers

"You are taking my personality away!"

"I feel like a first-year teacher."

"Everything about me is different this year."

Scientists, teacher associations, and government agencies are advocating a move from didactic information-centered science instruction to a more process-oriented approach (Rutherford & Ahlgren, 1990). Inquiry has been identified as a "critical component of the science curriculum" (National Research Council, 1993, p. 4). Constructivism[1] and authentic assessment are proposed to provide the foundational base and evaluative framework for that new curriculum (Sivertsen, 1993). These changes should, theoretically, lead to improved classroom environments and enhanced student understanding of the content and nature of science.

The prescribed changes will also require dramatic transformations in the basic beliefs and practices of many science teachers. The personal and professional upheavals that result from significant alterations in pedagogical practice are illustrated by the statements made at the beginning of this chapter. These

were taken from conversations with Colleen (pseudonym), a teacher involved in a curriculum implementation project. Colleen had been teaching high school science for nine years and had been very successful by her school district's standards—good test scores, high success rates, few parent complaints. She voluntarily agreed to abandon her preferred textbook-based method to implement a curricular model paralleling proposed state and national standards. Her comments remind us that methodological change involves more than altering lesson plans and changing textbooks. Colleen gave up many personally rewarding aspects of teaching. The class rules, usually posted before the school year began, were now negotiated by the students during the first week of classes. She stopped telling her students what they were to think about, what they were to do, or what the immediate answers to their questions were. She began to use her planning periods and evenings trying out laboratory activities and writing reflections in her journal. She gave fewer exams, added performance assessments, began assigning partial credit for assignments, and allowed students to redo unsatisfactory work. All these *minor* changes in Colleen's daily professional practice resulted in profound restructuring. It is easy to understand Colleen's

National Science Teachers Association

personal assessment that, "Everything about me is different this year."

The process of reform cannot be oversimplified; one shot show-and-tell workshops and fix-it-now attitudes will not produce long-term, productive changes in science classrooms. Political and social issues will need to be addressed. The deep-seated, philosophical beliefs about teaching and learning and the impact of those beliefs on established practice must be recognized by teachers and administrators. The title of this chapter reflects a belief that past efforts to address change in schools imitated the infomercials that invade television programming. Thus, instead of being handled as the sensitive issue it is, change is treated like a next-century car polish or the latest diet fad. Hand waving and quick-fix remedies will not facilitate a movement from science-as-information to science-as-process.

This chapter represents a synthesis of what the authors have found to be helpful in encouraging teachers to reflect upon and change science classroom practice. There is no one particular way to facilitate change, but support personnel (i.e., supervisors, university faculty, and administrators) can play pivotal roles in the creation of environments conducive to change.

## EMPOWERING CHANGE

Change is often perceived as an admission that we haven't been doing a particular thing very well. When change is viewed that way, the person asked to make a change will likely be defensive and self-protective. Risk taking is out of the question. It seems particularly important to help teachers consider change as a positive, natural part of professional development—a process over which they have control. Barrow and Tobin (1993) contend that "reform tends to be something done to someone else" (p. 115). The classroom teacher can easily become a victim of reform rather than a participant

in it. Given the proper support and incentive, however, teachers can become an empowered, active group in the process of change in science education. Those responsible for assisting teachers in the change process may incorrectly perceive empowerment as a messy political agenda. Alternately, empowerment could be viewed as the process of recognizing value and acting on ability and potential. That notion would more likely inspire personal reflection and make professional growth possible.

In order to feel empowered to make changes, teachers need opportunities to explore alternatives, to make mistakes, and to reflect on their own experiences, beliefs, and roles without fear of negative judgments or actions. Participants in teacher enhancement projects can engage in a variety of activities that facilitate and encourage professional explorations. Surveys designed to give insight into teachers' beliefs and practices (e.g., the "Teacher's Biography, Attitude, and Belief Survey[2]") can initiate early discussions about reform issues. Activities that encourage teachers to write and reflect about metaphors for teaching can be used to assist the teachers in understanding their personal philosophy (e.g., "Teacher Metaphor and Professional Skills Survey"). Concept mapping (Novak & Gowin, 1984) activities such as "Me, the Science Teacher" can be particularly illuminating. When constructing the maps, teachers often include roles and activities that describe their professional life. "Counselor," "mom," "dad," "coach," and "nurse" commonly appear on the maps, but titles referring to pedagogical or content-related responsibilities are often noticeably absent. This could indicate that teachers are often rewarded more for "coaching" and "nursing" and "parenting" than they are for developing and facilitating learning environments. It is important for support personnel to value and reward efforts teachers make to improve classroom environments and instruction. Once those aspects of teaching are valued, teachers may be more willing to take risks and make changes. It is difficult to conceive of taking personal and professional risks if one does not believe value is being assigned to the thing being changed. This attitude was graphically summed up by a disgruntled teacher who asked, "Why stick your neck out when no one cares about teaching anyway?"

### FACILITATING CHANGE

As teachers begin to value themselves professionally, they will require effective models for methodological change, time to meet with other teachers, and opportunities to experiment in their own classrooms. The support personnel can then act as facilitators of change by providing for these needs.

## Exploring Models and Frameworks

The proposed reform agenda attacks almost every aspect of traditional, didactic instruction. Group-oriented learning replaces individualized instruction; student explorations replace teacher lectures; process-oriented assessments replace multiple-choice tests; student autonomy replaces teacher-dominated environments. (For an excellent teacher-generated list see Davis, Shaw, & McCarty, 1993.) It is an overwhelming list to ponder and to implement. Alternative notions of teaching and learning cannot be rammed through the system, but those ideas can be modeled, discussed, tried, and refined. Teachers are products of their own school experiences where science content was lectured and labs were verifications of professor-given information. Similarly, many seminars and workshops designed to encourage change fall back on traditional strategies rather than utilize student-centered methodology. For example, during a seminar to inform university faculty and science education leadership about constructivism, a well-known speaker gave a lengthy lecture on the topic. It would seem more logical for workshop facilitators to model alter-

native frameworks for teachers.[3] Consider a workshop designed to introduce teachers to the learning cycle as a way of implementing student-centered science instruction. The learning cycle (Karplus & Thier, 1967; Lawson, Abraham, & Renner, 1989), is an experience-based model that seems particularly useful for addressing the numerous mandates emanating from state and national agencies. It would not be appropriate to lecture to teachers about a method designed to parallel the inquiry nature of science. Teachers could instead participate in several learning cycle investigations. (This strengthens teachers' content understanding as well.) Ensuing discussions would then focus on student and teacher roles and expectations, differences between the learning cycle and didactic models, and assessment issues.

Alternative philosophical systems can also be shared with teachers. One such alternative notion is the "teacher as learner" metaphor. Many teachers claim personal metaphors similar to "teacher as captain of the ship." Conversations with those teachers indicate that the metaphor conveys a belief that they are responsible for what students learn. The "teacher as learner" metaphor, however, shifts the focus from the teacher as a responsible agent of learning to being an active participant in the learning process. This metaphor has powerful implications, especially in classroom settings where the teacher is concerned about making content errors and shies from methods that leave him/her uncertain of knowing the right answers.

## Opportunities to Explore

Once alternative methodologies and frameworks have been experienced, teachers need time to implement changes and to reflect on the implications of those changes. Teachers who attend a one day, district-mandated workshop are unlikely to leave ready to change both practice and perspective, but a series of workshops across an academic year can have an immense impact. In school districts where time is set aside for professional development, one day each month could be used to provide opportunities for teachers to explore alternative methods. The workshop facilitators could then request that teachers implement those methods in the classroom, record the lesson (audio or video), and review the recording. This assignment between workshop sessions can provide valuable insights for teachers and productive discussions in succeeding workshops. One of the most common comments teachers make after such an assignment is, "I learned that the students really can figure things out for themselves. I didn't even have to tell them to…(use a procedure)." That insight will be sufficient to encourage many teachers to try other innovations. Telling teachers that a new method will have a particular result has little impact; their own experiences will be more persuasive.

## SUPPORTING CHANGE

Finally, the supervisor must support change. That support may take many forms: encouraging schools to reward change, helping teachers make professional presentations, or battling restrictive policies. Above all, support means that the supervisor is an advocate more for the teachers and students than for the desired changes.

## Go-between

Intervention will generally give better results than interference, and teachers need a "go-between" when engaged in risk-taking activities. It is important that someone not only recognizes teachers' needs and fears, but also helps remove bureaucratic roadblocks and provide essential materials. This point may be illustrated by the authors' work with a teacher previously referred to as "John" (Westbrook & Rogers, 1994). Asked by a local school district to work with John, the authors were amazed at the condition of his high school science classroom. His stu-

dents sat in desks with attached, slanted tops. The classroom had no water source, sink, or tables, and was bereft of even the simplest materials with which to do experiments. The school's principal was immediately contacted and a request was made for tables to be brought into the room. Once tables replaced the slant-top desks, John began to increase the number of activities he allowed his students to do. This one act did not cause John to become a more effective teacher, but it did alter his attitude about the administration and his own place within the school community.

## Counteracting Fear

Fear is a self-protective, often immobilizing, emotion. Teachers are legitimately afraid of engaging in activities that might cause them to lose their jobs. Productive change can occur only in a supportive, non-threatening environment. When possible, support personnel can contact local and state agencies and request exceptions to superfluous mandates. Teachers are more likely to embrace change when they don't fear having their professional knuckles rapped during the process.

In several states, for example, high school teachers responsible for core content courses are required to give state-written, end-of-course exams at the end of each academic year. Results of student performance on these exams are often printed in local newspapers and used to evaluate teacher performance. Teachers involved in these testing ventures contend the exams prevent them from implementing innovative teaching methods (Westbrook, 1994). Although research reports (e.g., Kyle, Bonnstetter, & Gadsden, 1988; Andrews, Huber, & Clark, 1994) indicate that test scores do not fall when inquiry-based methods are implemented, teachers continue to fear that change will have punitive results. To support change efforts, some school districts request testing alternatives for teachers implementing innovative curricula.

Teachers also credit restrictive evaluation practices with perpetuating didactic practices. Principals often assert the importance of new methods, then criticize noise levels and alternative practices. Teachers contend that evaluators seem more concerned with checking off line items on a standard evaluation form than noticing concept development and scientific inquiry. Many evaluators are unfamiliar with alternate strategies for teacher evaluation. Non-tenured teachers are justifiably hesitant to implement new instructional formats when they believe administrators will not be supportive. Support personnel can assist the process by conducting informative workshops to allow principals and other evaluators to better understand the practical implications of the suggested reforms in science education.

## Sweat the Small Stuff

The obvious is often overlooked. As in Colleen's case, several small issues can result in tremendous overall change. The effective supervisor can act as a constructive diagnostician—helping teachers find ways to make better use of time, develop new management and assessment techniques, and write curricula to achieve instructional goals. Planning and preparation requirements, for example, change dramatically when a teacher begins using an inquiry-oriented approach. Materials must be located and gathered. Experiments have to be conducted to give the teacher a feel for possible questions, problems, and insights the students might have. However, many teachers do not use their planning periods for preparation. Activities and surveys designed to help teachers analyze their use of planning times and concrete suggestions for restructuring daily schedules could encourage many teachers to examine more time-consuming instructional alternatives.

## CONCLUSIONS

The notion that supervisors, administrators, and university faculty have to empower, facilitate, and support change is not particularly original, but that model has not yet become the predominant way of thinking about assisting the professional growth and development of teachers. The notion that telling is teaching still pervades administrative mindsets; change continues to be pursued through one-day workshops. As value is placed on what happens in the classroom, and time is provided for exploration and growth, teachers may begin to view professional change as a positive and rewarding venture.

## Footnotes

[1] Constructivism is the notion that we construct knowledge through experiences with our surroundings and with other people. Ernst von Glasersfeld (1987) has suggested that this view of knowledge shifts "the emphasis from the student's correct replication of what the teacher does, to the student's successful organization of his or her *own* experience" (p. 6).

[2] All surveys and questionnaires designed by the authors are available at no charge for use by interested parties.

[3] We have used the following activity to initiate teachers' reflections about constructivism. The teachers are first placed in small groups of three to five members. Each group receives a packet of building materials (for example, Tinker Toys™, Lincoln Logs™ and Legos™) and is asked to build a "house." Once the houses have been constructed, the groups share about how they built their houses. The teachers find they have constructed different structures because of the initial materials and their collective experience. After a thoughtful discussion of the activity, the teachers have an understanding of constructivism as a view that knowledge is constructed from experience.

National Science Teachers Association

# References

Andrews, D., Huber, R.A., & Clark, R. (1994). Hands-on, inquiry-based science and state and national testing. *North Carolina Science Teachers Association Journal, 3*, 17–20.

Barrow, D., & Tobin, K. (1993). Reflections on the role of teacher education in science curriculum reform. In *Excellence in educating teachers of science*, by P.A. Rubba, L.M. Campbell, & T.M. Dana (Eds.), pp. 115–130. Columbus, OH: ERIC.

Davis, N.T., Shaw, K.L., & McCarty, B. J. (1993). Creating cultures for change in mathematics and science teaching. In *Excellence in educating teachers of science*, P.A. Rubba, L.M. Campbell, & T.M. Dana (Eds.), pp. 221–235. Columbus, OH: ERIC.

Karplus, R., & Thier, H.D. (1967). *A new way to look at elementary school science*. Chicago: Rand McNally.

Kyle, W.C., Jr., Bonnstetter, R.J., & Gadsden, T., Jr. (1988). An implementation study: An analysis of elementary students' and teachers' attitudes toward science in process-approach vs. traditional science classes. *Journal of Research in Science Teaching, 25*, 103–120.

Lawson, A.E., Abraham, M.R., & Renner, J.W. (1989). *A theory of instruction: Using the learning cycle to teach science concepts and thinking skills* (NARST Monograph No. 1). National Association for Research in Science Teaching.

National Research Council. (1996). *National science education standards*. Washington, D.C.: National Academy Press.

Novak, J.D., & Gowin, D.B. (1984). *Learning how to learn*. Cambridge, England: Cambridge University Press.

Rutherford, E.J., & Ahlgren, A. (1990). *Science for all Americans*. New York: Oxford University Press.

Sivertsen, M.L. (1993). *Transforming ideas for teaching and learning science*. Washington, D.C: U.S. Government Printing Office.

von Glasersfeld, E. (1987). Learning as a constructivist activity. In *Problems of representation in the teaching and learning of mathematics*, by C. Janvier (Ed.), pp. 3–17. Hillsdale, NJ: Erlbaum.

Westbrook, S.L. (1994). [Responses to Teacher Biography, Attitudes, and Beliefs Survey.] Unpublished raw data.

Westbrook, S.L., & Rogers, L.N. (March, 1994). *Implementing reform in rural high school science classrooms: A case study of two physical science teachers*. Paper presented at the annual meeting of the National Association for Research in Science Teaching in Anaheim, CA.

## Author Note

**Laura N. Rogers** is an assistant professor of science education at the Watson School of Education, University of North Carolina at Wilmington. Dr. Rogers is interested in the development of collaborative learning environments that facilitate teachers' reflective practice and students' autonomous constructions of science concepts.

**Susan L. Westbrook** is an assistant professor of science education in North Carolina State University's Department of Mathematics, Science, and Technology Education. Dr. Westbrook has previously developed, taught, and researched learning cycle science curricula and is currently exploring ways to integrate mathematics and science content using that model.

# How Standards Fit within the Framework of Science Education Reform

Andrew Ahlgren

Several organizations have taken a national lead in fostering reform in science, mathematics, and technology education, including the development of standards. In 1989, Project 2061 of the American Association for the Advancement of Science published *Science for All Americans* (*SFAA*), which recommended a coherent set of learning goals in science, mathematics, and technology for all high school graduates. That same year, the National Council of Teachers of Mathematics released their *Curriculum and Evaluation Standards for School Mathematics*, and the National Science Teachers Association initiated a variety of alternative scope, sequence, and coordination curriculum-design projects.

Since then, Project 2061 has gone on to publish *Benchmarks for Science Literacy* (1993), which describes what students should know and be able to do in science, mathematics, and technology as they progress through 13 years of schooling. And in December 1995, the National Academy of Sciences' National Research Council released its *National Science Education Standards* for grades 4, 8, and 12. Other important players include the New Standards Assessment Project at Carnegie-Mellon University and the National Assessment of Edu-

cational Progress in Princeton, both of which will attempt to make their work consistent with national guidelines for science education.

## MAKING A DIFFERENCE

Together, these reform efforts represent the work of thousands of individual scientists and educators and hundreds of scientific and educational organizations. Believing that the nation would benefit most from a consensus within the national leadership, all those involved are working toward consistency among their products and efforts. This kind of agreement will give educators the confidence and freedom to make use of any—or all—of them as they choose.

These formal, national efforts cap more than a decade of informal progress in rethinking science education, begun in 1980 with the publication of *Project Synthesis* (Harms, 1980). The most recent data (1992) from the National Assessment of Educational Progress indicate improved student performance by 9-, 13-, and 17-year-olds during that period:

Educating America's youth to become scientifically literate has become a primary goal for our nation in recent years and methods for improving science education are major topics of discussion in the classroom, school, and district as evidenced by the presentations and workshops being given around the country. This willingness of educators to provide a forum for discussion on how to improve science education may be paying dividends, as evidenced by the significantly increased proficiency between 1982 and 1992 at all three ages. (p. 33)

A number of states—Alabama, Georgia, Maryland, Michigan, and Ohio, for example—have based their new or revised science curriculum frameworks on the science literacy goals recommended in Project 2061's *Science for All Americans*. As other states, urban centers, and rural areas begin to reform their school systems under the National Science Foundation's Systemic Initiative programs, they too will be setting clear goals for what students should know and be able to do and will likely be turning to the national reform projects for guidance.

Because Project 2061's vision of long-term systemic reform of the science education system was one of the first to gain national acceptance, this paper will focus particular attention on its work. It will also look at how the science literacy goals expressed in *SFAA* and *Benchmarks* have contributed to the development of other goals-based science education reform efforts.

## SETTING GOALS

The move toward national standards throughout education is the result, in large part, of the failure of piecemeal, sporadic, and disconnected reform efforts of the past to bring about needed change. While there were isolated instances of success, they had been hard to transfer from one classroom or school to another, let alone from one district or state to another. None seemed able to bring about lasting change. Most were unfocused in their

aims. By 1985, when Project 2061 was launched, science educators were calling for consensus among educators, scientists, and publishers on goals for science education, on effective teaching methods, and on content. What was needed was a broad vision of what it means to be science literate.

With the publication of *SFAA*, Project 2061 proposed a new way of thinking about how to educate and equip students for a world that is increasingly shaped by science, mathematics, and technology. It defines science literacy in this way:

… being familiar with the natural world and respecting its unity; being aware of some of the important ways in which mathematics, technology, and the sciences depend upon one another; understanding some of the key concepts and principles of science; having a capacity for scientific ways of thinking; knowing that science, mathematics, and technology are human enterprises, and knowing what that implies about their strengths and limitations; and being able to use scientific knowledge and ways of thinking for personal and social purposes. (AAAS, 1989, p. xvii)

The basic proposition, widely subscribed to by science educators, was that there was far too much for students to learn meaningfully in the "overstuffed and undernourished" science curriculum. Choices would have to be made about what knowledge and skills offered the most payoff—for understanding how the world works and for learning more—and therefore needed to be learned well and retained. *SFAA* recommends a specific, interconnected, and coherent set of learning goals in science, mathematics, and technology for all high school graduates:

---

- Chapters 1 through 3 focus on the nature of science, mathematics, and technology as human enterprises.

- Chapters 4 through 9 cover basic knowledge about the world as currently seen from the perspective of science and mathematics and as shaped by technology.

- Chapters 10 and 11 present what people should understand about some of the momentous episodes in the history of the scientific endeavor and about some crosscutting themes that can serve as tools for thinking about how the world works.

- Chapter 12 lays out the habits of mind—attitudes, skills, and perspectives—that are essential for science literacy.

SFAA also makes a strong case for education reform, outlines an agenda for action, and offers some principles for effective teaching and learning.

The 1989 publication of SFAA defined for the nation what science literacy should encompass and what students would need to know in science, mathematics, and technology to lead responsible and interesting lives in the 21st century. It also provided additional impetus for the national standards movement. By 1990, the U.S. Department of Education (at the urging of the National Science Teachers Association) made the National Academy of Sciences responsible for coordinating the development of science education standards and for building the necessary political support for their adoption.

## A CONSENSUS ON CONTENT

Project 2061 continued its work with scientists and teams of educators at its six school-district centers to translate the literacy goals described in SFAA into specific learning goals for students as they progressed through school. The result of this effort was a set of statements, or benchmarks, about what students should know and be able to do by the end of grades 2, 5, 8, and 12 as they progress toward the outcomes expected of high school graduates presented in SFAA. (The additional step—grades 2 and 5 instead of grade 4—was introduced on the advice of the school-district centers that qualitatively different expectations were appropriate for primary school.)   Further, the new document included essays to provide context for the statements and to relate them to Project 2061's broader reform vision. An extensive compilation of education research material to support the substance and grade placement of benchmarks was also included. The document was reviewed by more than 1,300 individuals—teachers, administrators, scientists, mathematicians, engineers, historians, and experts on learning and curriculum design—in diverse school districts in 46 states.

Benchmarks for Science Literacy was published in late 1993. The response from educators was extremely positive. To many, Project 2061's benchmarks provided the long-awaited guidance they needed to begin the task of science education reform.

With both SFAA and Benchmarks available to guide them, educators were encouraged to explore in detail the concept of science literacy and its implications for instruction at all levels. States, school districts, developers of instructional materials, teacher educators, and classroom teachers themselves began to develop or revise their curriculum frameworks, evaluate existing instructional materials to see how well they contribute toward science literacy, and consider how best to assess student learning. At the same time, the work of Project 2061 was being incorporated into the proposals and short- and long-term reform plans of many of the National Science Foundation's State, Urban, and Rural Systemic Initiative grantees. Currently, Project 2061 is developing a number of additional print and electronic tools to help educators improve curriculum and instruction—including print and disk versions of *Resources for Science Literacy: Professional Development,* which presents flexible guides for a variety of workshops on how to use benchmarks.

Project 2061's work has also helped to shape the *National Science Education Standards* developed by the National Research Council (the operating arm of the National Academy of Sciences). Project 2061 has shared with them its own work on *Benchmarks*, first in draft form and then as published books. Project 2061 staff have conferred with the *Standards* writing staff and have served on its various working committees.

The *Standards* provides, in a single volume, standards in six areas of science education: content, teaching, professional development, assessment, program, and systems. Project 2061, on the other hand, has focused mainly on content in *SFAA* and *Benchmarks.* In its work-in-progress *Blueprints for Reform,* however, Project 2061 has been taking a long look at the systemic aspects of reform, integrating the findings of 12 expert groups it has commissioned to examine how various parts of the education system—higher education, materials and technology, school organization, research, family and community, and business and industry, among others—should change to accommodate Project 2061's curriculum and teaching reforms.

Unlike Project 2061's *Benchmarks,* which recommends learning goals in mathematics, technology, and social science along with natural science, the content portion of the *Standards* deals mostly with natural science. Aside from this difference, there is a high level of congruence between the content portion of the *Standards* and the work of Project 2061, principally *SFAA* and *Benchmarks,* as indicated in a detailed analysis of both sets of recommendations conducted by Project 2061 with the cooperation of the

NRC (Project 2061, 1995). For example, the two sets of recommendations largely agree on what is essential for all students to know—and what is not. They exclude from the core curriculum many of the same topics that clutter the current science curriculum: Ohm's law, series and parallel circuits, geometric optics, phyla of plants and animals, cloud types, simple machines, and many others. They present their recommendations similarly, focusing on real understanding of important facts and ideas, greatly de-emphasizing vocabulary memorization that too often substitutes for understanding, and placing at specific grade levels the statements of ideas and concepts students should learn. The "fundamental concepts" presented in the *Standards* and Project 2061's *Benchmarks* are close in their scope, level of difficulty, and grade placement. (The greatest current difference is that the *Standards* recommends that science literacy for all should include the ability—as adults—to actually carry out scientific investigation, in addition to understanding it. Where there are still differences in detail or emphasis, revised versions are expected to reduce them. And the conversation will continue; neither AAAS or the National Research Council claim that their products are complete, perfect, or immutable. The NRS has just recently installed a Center for Science, Mathematics, and Engineering Education, which will keep an eye on how it is used.

The National Science Teachers Association (NSTA), which was instrumental in assigning to the NRC the responsibility for developing standards, has been disseminating *SFAA* and *Benchmarks,* has already begun to promulgate the *Standards,* the print and disk resources of *Pathways to the Science Standards,* Awareness Kits, and Training Institutes. As an example of how national standards might be applied in designing alternative science curricula, the NSTA drew from the *Standards* for its own scope, sequence, and coordination curriculum designs. Several versions of this new curriculum approach seek to replace the traditional "layer-cake" science courses with a coordinated curriculum in which students study the physical, biological, and earth and space sciences, in parallel, every year from junior high school through senior high school.

There are a great many other curriculum-design projects, each tailoring national advice to their own circumstances and creativity. Change of this magnitude can be difficult, and one of the biggest dangers is that curriculum designers at any level might misunderstand, or even dismiss, the hard won consensus on reducing the amount of material students must learn and, instead, continue to include most of the current content, merely re-sorting it so that it appears to fit specific standards.

### A MEANS, NOT AN END

No matter how skillfully crafted, standards alone cannot reform education. The core reform message—of uncluttering the curriculum so that what is learned is meaningful, of including *all* students, and of addressing many aspects of the system at the same time—is going to meet vigorous resistance. As was stated in a recent article in *The New Yorker*, serious reform "is profoundly threatening ... to people whose lives and outlooks have been shaped by the conventional system." (Traub, 1995, p. 78)

Leaders in science education have an unprecedented opportunity for reform, but much more will be needed. If standards are all that they should be, then many schools will find themselves falling short. These schools will need financial, human, and curricular resources to meet the standards. New teaching methods, materials, and assessment tools must be developed so that standards can be implemented. All of this requires a commitment far beyond the adoption of standards. In many ways, the hardest work still lies ahead.

## References

American Association for the Advancement of Science. (1989). *Science for all Americans.* New York: Oxford University Press.

American Association for the Advancement of Science. (1993). *Benchmarks for science literacy,* New York: Oxford University Press.

Directorate for Education and Human Resources. (1994). *Foundation for the future: The systemic cornerstone,* Washington, DC: National Science Foundation.

National Assessment of Educational Progress. (1994). *1992 science trend assessment.* Princeton, NJ: Author.

National Council of Teachers of Mathematics. (1989) *Curriculum and evaluation standards for school mathematics.* Reston, VA: Author.

National Research Council. (1996). *National science education standards.* Washington, D.C.: National Academy Press.

Project on Scope, Sequence, and Coordination of Secondary School Science. (1993). *The content core.* Vol. 1. Washington, D.C.: National Science Teachers Association.

Project on Scope, Sequence, and Coordination of Secondary School Science. (1995). *A high school framework for national science education standards.* Vol. 3. Arlington, VA: National Science Teachers Association.

Project 2061. (1995) *Comparison of content between the November draft of National Science Education Standards and Benchmarks for Science Literacy/Science for All Americans.* (available from Project 2061, American Association for the Advancement of Science, 1333 H Street, N.W., Washington, D.C. 20005).

Traub, James. (1995, July 17). It's elementary. *The New Yorker.* 74–78.

## Author Note

**Andrew Ahlgren** is associate director of Project 2061 at the American Association for the Advancement of Science and professor emeritus of curriculum and instruction at the University of Minnesota. He is a principal contributor to *Science for All Americans* and *Benchmarks for Science Literacy*.

# NSTA's Pathways Project: Using the
## *National Science Education Standards*

Shirley Watt Ireton

Until now, we have not had true standards for science education. Educators have followed de facto standards: teaching the information needed for good scores on standardized tests, or following the exhaustive content of science textbooks. The 263-page report, *National Science Education Standards*, released by the National Research Council (NRC) at the end of 1995, provides a new look at the many needs and goals of science education. This article describes the advent of the *Standards*, and the projects being developed by the National Science Teachers Association (NSTA) to facilitate their use.

Development of the *Standards* began in the spring of 1991, when NSTA—joined by scientific societies, other science education organizations, the U.S. secretary of education, the National Science Foundation's assistant director for education and human resources, and others—formally asked the NRC to coordinate development of national science education standards. The need for standards had become evident—pulled in many directions, science educators, administrators, and researchers sorely needed a unifying plan to coordinate their work. Individual, unconnected efforts to fill the need for standards were beginning, but each

served a niche, and links or interaction among them was difficult. Consensus settled on the NRC as the group who would link the many factions of science education. Funding was forthcoming from the Department of Education and the National Science Foundation.

Inclusion of the word "National" in the title of the *Standards* derives from the vast input and review by groups and individuals throughout the country, sought and included by the NRC as part of the four-year development process. The Advisory Committee members realized that without input from all groups with expertise and interest in science education, the *Standards* would not truly be standards. The committee therefore included representatives from science education, science supervision, and scientific associations. The American Association for the Advancement of Science's *Project 2061*, NSTA's *Scope Sequence & Coordination* project, and other seminal projects were drawn from. As the pre-draft was readied, focus groups from all the stake holders critiqued the document. At draft stage, more than 40,000 copies were sent for review by more than 18,000 individuals and 250 groups. The final version incorporates these views. It is rare when a profession can achieve national consensus in its vision for change. The *National Science Education Standards* are truly a consensus among science educators and others of what is important in a child's K–12 science education experience.

## THE INTENT OF THE STANDARDS

From the beginning of development of the *Standards*, participants recognized that science content, though most obvious, was but a component of science education. The *Standards* are presented in six chapters: about science teaching, professional development, assessment, science content, science education programs, and science education systems. They offer what students need to know, understand, and be able to do to be scientifically literate. They describe an educational system in which "all students demonstrate high levels of performance, in which teachers are empowered to make the decisions essential for effective learning, in which interlocking communities of teachers and students are focused on learning science, and in which supportive educational programs and systems nurture achievement." ( NRC, 1996)

The *Standards* show an ideal state—a destination that may seem distant to busy teachers in schools across the United States. In classrooms where the staff has taken on many social responsibilities apart from education, science teachers may find it difficult to read the *Standards* and determine how to begin to implement them. The very breadth of this landmark document may seem at first intimidating—something for everyone, but nothing for tomorrow's lesson plan.

The *Standards* are fundamentally practical. According to educator and curriculum developer Karen Worth, "There is nothing in the new *Standards* that a really good teacher in the right environment is not doing already." There is the challenge. Teachers don't always have "the right environment" for teaching. They are asked to teach outside their areas of expertise, to teach without sufficient preparation time. The task for teachers is to identify aspects of their work and their environment that already fit the *Standards* recommendations, and then evaluate paths for change.

The *Standards* outline science education; they don't describe how to effect change. NSTA's *Pathways* project is designed to provide that guidance.

## THE NSTA PATHWAYS PROJECT

*Pathways to the Science Standards: Guidelines from Moving the Vision into Practice* are practical tools for science teachers to use to move their teaching, professional development, assessment, program and curriculum, and interactions with the larger education system toward the vision of the *Standards*. The authors are NSTA's members—educators, curriculum developers, and administrators—who have drawn from their experience to provide guidance for their colleagues.

*Pathways* begins with three practical volumes—for elementary, middle level, and high school—that provide specific suggestions and clear examples for science teachers to use to implement each of the standards. Resources for the Road—articles which provide further explanations and suggestions—accompany each chapter. Most of these resources are on the Mac/Windows CD included with each book. Each chapter also diagrams options for growth to fit the varied needs of schools and classes. These diagrams include landmarks schools and classes might use when aiming for a destination somewhat closer to the *Standards* than they are today.

There is no linear progression from "bad" to "good" in attempting to move the vision of the *Standards* into reality. The *Pathways* are designed to guide teachers to evaluate their existing programs, and to then evaluate paths of change to improve their programs. There are as many paths to change as there are modes of learning in our students.

The *Pathways* books begin by building on what works well in classrooms and schools today. That hasn't been difficult to find. Pilot projects and established programs that clearly illustrate the vision of excellence are found today around the nation. These projects have been described in NSTA's journals, presented at conventions, found acclaim in NSTA's projects and programs. The books mine such resources, and others, to provide examples of classrooms exemplifying the *Standards*.

The *Pathways* chapters devoted to the science content outlined in the *Standards* provide guidance on evaluating existing content in terms of the standards, and on adapting teaching methods. For example, in high school life science, the study of the cell is the first content area defined, and traditionally the first presented. The challenge for teachers is the bridge between what can be seen (and what concrete thinkers can understand) and the microscopic relationships between structure and function. In a *Standards*-based classroom, macroscopic investigations (such as the infusion of *Elodea* with saline solution) will be paired with microscopic observations (of cell lysis) in order to help students bridge the gap for concrete to formal logic. At the same time, teachers will avoid allowing students to rely on rote learning (such as name memorization)to achieve success.

While the casual reader of the *Standards* may find the content daunting, it represents a far slimmer core of knowledge than most science courses would previously have required. "Learn the core ideas, and learn them well" is the message of the *Standards*. The implied corollary is, "Do what is most important well, and leave the rest."

## NSTA'S COMMITMENT TO THE STANDARDS

As an organization of science education professionals, with a stated purpose of stimulating, improving, and coordinating science teaching and learning, NSTA has embraced the *Standards* and made a commitment to their further development. NSTA publications will now have *Standards* matrices to guide teachers in using NSTA's publications and journal articles. *Guidelines for Self-Assessment*, the premier program evaluation tool for science educators, is being redesigned to reflect accomplishment of goals versus desirability of goals in terms of the standards. Conventions and workshops will reflect the ideas of this new document. Perhaps the many "crises" that have been trumpeted about science education over recent decades will have found a pathway to solution.

## References

American Association for the Advancement of Science, Project 2061. (1989). *Science for All Americans*. New York: Oxford University Press.

National Research Council. (1996). *National Science Education Standards*. DC: National Academy Press.

National Science Teachers Association. (1996). *Pathways to the Science Standards: Guidelines from Moving the Vision into Practice*. Arlington, VA: Author.

## Author Note

**Shirley Watt Ireton** is NSTA's assistant director of publications and managing editor of special publications. Education at Purdue in microbiology, then in journalism at the University of Maryland led her to NSTA. For the past 14 years she has developed and managed science education projects and publications, including numerous award-winning publications. Her particular interests are conveying leading edge science to educators in classroom-ready ways, and modeling innovative teaching techniques in the development of science education publications.

# Technology and the Science Program: Placing Them in the Proper Perspective

Ronald E. Converse

Science educators should be proud of the various efforts to reform science education. In reviewing publications of the Association for Supervision and Curriculum Development (ASCD) over the last two years, it is evident that reform efforts in science education are utilizing the most current thinking in curriculum design. One of the latest ASCD publications (Beane, 1995), for instance, cites the importance of developing a coherent curriculum that utilizes "big" ideas to hold the curriculum together and provide a more relevant program than students encounter today. One example of efforts to develop a coherent curriculum cited in this publication is Project 2061 (Ahlgren & Kesidou, 1995).

For some time, a wide variety of uses of technology has been described in various publications. For example, Solomon (1994) described a number of examples such as promoting authentic research, utilizing networks for competitions, and interdisciplinary testing. Johnston (1995) described a variety of uses of technology in the physics classroom. Among these are microcomputer-based laboratories, simulations, spreadsheets, computer-based tests for individuals, and cooperative groups. The use of spreadsheet templates in chemistry at the college level is described by Joshi (1993). An interesting project, the Technology, Science, Mathematics (TSM) Integration Project, has been reported on by Sanders (1994). The majority of these articles are interesting and provide ideas that might be incorporated into programs elsewhere.

However, such articles are not meant to provide a coherent view of technology and its role in efforts to reform science education. As leaders in the effort to improve science education, science supervisors need to understand technology and its relationship to curriculum reform. The majority of persons responsible for providing leadership in science education have paid close attention to the major reform movements in curriculum and instruction, in general, and science in particular. However, many in educational leadership positions do not understand the relationship of technology to efforts toward reforming and restructuring education.

## DEFINITION

In general, technology is defined as applied science. Today, most people commonly classify a number of electronic devices that pervade our society as technology. Among these devices are computers, laser discs, VCRs, and television. At various times in the past, a pencil, fountain pen, film projector, opaque projector, overhead projector, 16-mm film loop projector, or cassette tape player/recorder would have come to mind as examples of technology. Those who have been in education for some time will recall the 16-mm film loop projectors. Those just beginning their education careers may not be familiar with this device. It is evident that none of the technological devices listed have thus far had a lasting positive influence on education. Snider (1992) suggests that the printing press today, as it has for centuries, remains as the predominate mechanical teaching aid in education.

## TWO MOVEMENTS

Thornburg (1992) indicates that there has been, in many cases, a societal transformation during the transition years between centuries. However, although great changes took place in society and its institutions during these transition periods, education has been relatively unaffected. It is the opinion of Thornburg that the pressures on public education at the close of the 20th century will force radical changes as we enter the 21st century.

Thornburg (1992) and Means (1994) both agree that there are two movements taking place in education as we move into the 21st century. The first of these is the effort to restructure education, while the second is the rapid introduction of technology into schools. Regarding these two movements, Means writes:

Unfortunately, these two innovative trends do not always (and some would argue do not usually) occur in concert. In many places, serious school reform is being undertaken without any real consideration of the facilitating role that technology might play. Even more common is the introduction of new instructional technologies without any serious consideration of how these

technologies might further school reform goals. (p. xi)

The above quotation states what should be one of the primary concerns of any person involved in exploring ways to reform science education. Kirkpatrick (1994) echoes the concern that the introduction of technology does not occur as part of the development of educational goals. Riel (1994) discusses many components involved in restructuring education, including collaborative technology that helps individuals produce shared knowledge. However, Riel writes that technical innovations alone will not reorganize education.

## THE PROPER ROLE OF TECHNOLOGY

A basic question needs to be addressed when striving to improve science education at the department, school, district, state, or national level. Are these efforts utilizing technology appropriately? Is a conscious effort being made to incorporate technology in a manner that will facilitate and enhance the desired change? Research by Bruder, Buchsbaum, Hill, and Orlando (1992) indicates that even in reform efforts where technology played a minor role, it was an essential part.

When working on improving science education, it is important to place the reform effort and technology in the proper perspective. During an examination of technology and the new science literacy, Bruder (1993) indicates that student educational goals must come first, then technology that facilitates student attainment of the goals is determined. This viewpoint is echoed by a number of other writers including Barron and Orwig (1993) along with Dede (1994). Whitaker (1995), describing how a district utilized technology to help improve instruction, supports the view that district goals were established first, then the role of technology to support those goals was developed. Herman (1994) states that: "Just as technology must be built on significant and meaningful curricula, so efforts to integrate technology into schools must be combined with professional development for teachers in effective curriculum design and instruction" (p. 164).

## RESEARCH ON USING TECHNOLOGY

Ritchie and Wiburg (1994) examined variables that influenced the integration of technology into the curriculum. The variables are consistent with other research regarding changes in education. The variables identified are (a) administrative leadership and support, (b) pedagogical orientation of the teacher, (c) staff development, and (d) collaborative partnerships with outside organizations. Schools in which technology was being used effectively had administrators who used technology on a consistent basis. These administrators were working toward restructuring and had effectively communicated a vision for reform to their faculty and community. The schools also had well developed technology plans that supported the restructuring efforts and empowered faculty to utilize technology. The administrators participated frequently in staff development or training related to technology. The majority of cases examined, where the integration of technology was successful, also had partnerships established with businesses, institutions of higher learning, or offices of education.

The examination by Ritchie and Wiburg of the second variable influencing technology integration, the pedagogical orientation of the teacher, is startling. Drawing from a number of research studies, they indicated that teachers with a high level of technology implementation shared a number of characteristics. These teachers:

- focused on students developing a sense of curiosity;

- instilled in their students a desire to learn;

- used technology as a tool for thinking and learning more deeply about content;

- reduced the time devoted to the acquisition of facts;

- devoted more time to an inquiry-based approach;

- assisted development of critical thinking skills;

- utilized a process approach to meeting curricular objectives;

- allowed students to utilize technology in creative ways to master curricular objectives; and

- placed more importance on how students approached solving problems than on test scores.

These are many of the teacher behaviors advocated in various movements to reform science education. Even more importantly, these characteristics parallel teacher behaviors found in reform efforts of other disciplines. However, it has been difficult to foster these characteristics in the majority of science teachers. Technology might increase the number of teachers displaying these characteristics and be a catalyst for fostering learning environments that integrate various disciplines. Increasingly, teachers have the ability to network with their peers outside the school. This collaboration among teachers, properly used, could increase the use of successful instructional practices.

Also interesting are the characteristics of teachers demonstrating a low level of technology implementation according to Ritchie and Wiburg (1994). These teachers were much more heterogeneous regarding their approach to teaching. While some were process orientated, as a group, teachers with a low level of technology implementation exhibited traditional approaches to teaching. These teachers:

- maintained structured classrooms;

- demonstrated high levels of discipline;

- emphasized content rather than processes;

- followed the textbook closely; and

- lectured as a primary means of instruction.

Basically, the characteristics of teachers demonstrating a low level of technology implementation are those that describe characteristics of the classroom for the last 100 years of education. Both groups of teachers, high users and low users of technology, had one thing in common. Both indicated they had not received sufficient training in the use of technology.

Khalili and Shashaani (1994), summarizing a meta-analysis of 151 reported studies on the effects of technology in education, indicated that the average effect size was .38. The highest effect sizes were found at the high school level and the lowest at the middle school level. It is important to note that this study indicated that as technology has become more sophisticated and available, its effectiveness in improving student learning has not necessarily increased. However, many of the studies included in this meta-analysis were evaluated using traditional testing procedures. They may not indicate the potential for technology to aid reform efforts in education. Khalili and Shashaani indicated that little research in this area examined how instruction was delivered in conjunction with technology.

Good and Brophy (1994, pp. 331–332), Herman (1994), and Reeves (1992) all suggested that additional research is needed on the use of computers as related to reform efforts in education. Reeves suggested that evaluation of restructured education must involve four functions. These functions are goal refinement, documentation, formative experimentation, and impact assessment. He then provided specific information on evaluation design and methods appropriate for evaluating restructured education. Herman did an excellent job of discussing the issues involved in evaluating the effects of technology in school reform. She provided several techniques that can be utilized in the evaluation process. Examples of successful applications are provided as well as an examination of the difficulties of determining evaluation procedures. One of the successful applications cited by Herman (1994) involves several studies indicating microcomputer-based laboratories (MBLs) assist students to understand scientific concepts.

## TECHNOLOGY IS NEUTRAL

Technology is neutral (Thornburg, 1992). It can be a positive force in supporting curriculum reform. On the other hand, technology can be used to entrench undesirable practices. Wiburg (1994) cautions that: "As advocates of responsible uses of technology we must ensure that our use aligns with our instructional intentions rather than undermining them" (p. 234). Too often we jump at the chance to add more technology to our science classrooms without serious consideration of the impact this addition might have on our efforts to improve instruction.

As supervisors, we often hear: "Just give us the technology, and then we'll decide how to use it." With some teachers who are already making strides in improving their classrooms, this might have merit, primarily because these teachers already have a clear vision of how they are going to improve science instruction for their students. They will use technology to facilitate the attainment of the vision. Unfortunately, for teachers who do not have a vision of how they wish to restructure their classrooms,

National Science Teachers Association

providing technology will make little or no difference. In fact, they may use technology to support practices that should be altered and in the process construct another barrier to reform efforts.

## Using Technology Effectively

If leaders in science education are to use technology effectively, they must make a conscious effort to educate themselves in this area. As cited earlier (Ritchie and Wiburg, 1994), where technology is being used effectively, administrators are avid users of technology. Not only should supervisors provide and encourage teachers to participate in staff development regarding technology, they should also participate enthusiastically.

It is also important to expand one's professional reading to include research in the use of technology in addition to the area of curriculum and instruction. The National School Boards Association has formed a collaborative Technology Leadership Network that provides assistance to school districts wishing to infuse technology. There are also several national and state professional associations related to technology. Dyrli and Kinnaman (1994) provided an excellent series of articles over a five month period that present a very good overview of the role of technology.

Most important to utilizing technology effectively is to first develop a clear, concise description of the reforms desired in science education. This description should be in alignment with national (West, 1994), state, and district goals. Then, determine where technology can best facilitate the attainment of the desired goals in science education. Peck and Dorricott (1994) listed ten reasons for using technology and described how technology facilitates each. While there are several different authors' views of where technology best facilitates instructional change, this list does provide an idea of the wide area where technology can be of

use. These ten reasons are:

1. Students learn and develop at different rates.

2. Graduates must be proficient at accessing, evaluating, and communicating information.

3. Technology can foster an increase in the quantity and quality of students' thinking and writing.

4. Graduates must solve complex problems.

5. Technology can nurture artistic expression.

6. Graduates must be globally aware and able to use resources that exist outside the school.

7. Technology creates opportunities for students to do meaningful work.

8. All students need access to high level and high-interest courses.

9. Students must feel comfortable with the tools of the Information Age.

10. Schools must increase their productivity and efficiency.

Areas where there is an alignment between the goals of desired science reform efforts and the areas where technology has been shown to be successful are logical places to infuse technology into the science program.

Sometimes, it is even more important to say "no thanks" when being offered technology for use in the classroom. If the technology offered doesn't facilitate the science education reform efforts, it could cause a distraction that takes teachers' efforts away from the primary objective. In a worst-case scenario, the technology could actually be a negative influence.

## KNOW WHERE TECHNOLOGY IS GOING

Science education leaders must keep looking to the future. Just as reform efforts evolve and change over time based on formative evaluation, technology also evolves due to a number of factors.

Communications technology is changing rapidly. Itzkan (1994), while examining the future of telecomputing environments, made the observation that the time required for a technology to move from business to education is rapidly shrinking. Rich et al. (1994) have described the use of virtual reality in collaborative activities among students. Rapid changes in communications technology can redefine the term "classroom." A classroom may become an electronic location where learners meet.

Normally, when persons talk about the future of technology, it is as if the microcomputer will go on forever. We will simply move from 286 to 386 to 486 to 586, ....., to super PC, and on and on. Why should we believe this? The technology of today is not the technology of yesterday. Certainly we can not expect the technology of tomorrow to be the predominant technology of today. Gilder (1994) described the relationship between network bandwidth and internal personal computer bandwidth. Network bandwidth is advancing ten times faster than central processors. In the next two decades, the predominant technology will be digital networks with mediaprocessors rather than microprocessors. Gilder states that: "In time the microprocessor will become a vestigial link to the legacy systems such as word processing and spreadsheets that once defined the machine" (p. 177).

## SUMMARY

Restructuring education in general and specifically science education must be the primary goal for leaders in science education. The infusion of technology is dependent on the extent to which it facilitates the desired changes in curriculum and instruction defined for science education. Technology can facilitate the realization of a vision of science education. But, like any tool, if used improperly, it can blur the vision and delay its realization.

## References

Ahlgren, A., & Kesidou, S. (1995). Attempting curriculum coherence in Project 2061. In *Toward a coherent curriculum*, by J. A. Beane (Ed.). Alexandria, VA: Association for Supervision and Curriculum Development.

Barron, A.E., & Orwig, G.W. (1993). *New technologies for education: A beginner's guide*. Englewood, CA: Libraries Unlimited.

Beane, J.A. (1995). Conclusion: Toward a coherent curriculum. *Toward a coherent curriculum*. Alexandria, VA: Association for Supervision and Curriculum Development.

Bruder, I. (1993). Redefining science. *Electronic Learning, 12*(6), 20–24, 29. In *Assessing the role of technology in education*, by Wishnietsky, D.H. (Ed.) (1994), Bloomington, IN: Phi Delta Kappa.

Bruder, I., Buchsbaum, H., Hill, M., & Orlando, L.C. (1992). School reform: Why you need technology to get there. *Electronic Learning, 11*(8), 22–23, 25–28. In *Assessing the role of technology in education*, by Wishnietsky, D.H. (Ed.) (1994). Bloomington, IN: Phi Delta Kappa.

Dede, C. (1994). Making the most of multimedia. *"Multimedia and learning: A school leader's guide"* by A.W. Ward (Ed.) (1994?). Alexandria, VA: National School Boards Association.

Dyrli, O.E., & Kinnaman, D.E. (1994). Gaining access to technology: First step in making a difference for your students. *Technology & Learning, 14*(4), 15–20, 48, 50.

Gilder, G. (1994, December 5). Telecosm: The bandwidth tidal wave. *Forbes ASAP: A Technology Supplement to Forbes Magazine*, pp. 162–167, 169–170, 172, 174, 176–177.

Good, T.L., & Brophy, J.E. (1994). *Looking in classrooms* (6th ed.). New York, NY: HarperCollins College.

Herman, J.L. (1994). Evaluating the effects of technology in school reform. In *Technology and education reform: The reality behind the promise*, by B. Means (Ed.) (1994?). San Francisco, CA: Jossey-Bass Publishers.

Itzkan, S.J. (1994). Assessing the future of telecomputing environments: Implications for instruction and administration. *The Computing Teacher, 22*(4), 60–67.

Johnston, M.A. (1995). The high-tech physics class. *The Executive Educator, 17*(2), 39–40, 47.

Joshi, B.D. (1993). Spreadsheet templates for chemical equilibrium calculations. *Journal of Computers in Mathematics and Science Teaching, 12*, 261–276.

Khalili, A., & Shashaani, L. (1994). The effectiveness of computer applications: A meta-analysis. *Journal of Research on Computing in Education, 27*, 48–61.

Kirkpatrick, C.M. (1994). Some thoughts on implementing technology. *ASCD Curriculum/Technology Quarterly, 4*(2), 6.

Means, B. (1994). Preface. In B. Means (Ed.), *Technology and education reform: The reality behind the promise*. San Francisco, CA: Jossey-Bass.

Peck, K.L., & Dorricott, D. (1994). Why Use Technology? *Educational Leadership, 51*(7), 11–14.

Reeves, T.C. (1992). Evaluating schools infused with technology. *Education and Urban Society, 24*(4), 519–534. In *Assessing the role of technology in education*, by Wishnietsky, D.H. (Ed.) (1994). Bloomington, IN: Phi Delta Kappa.

Rich, C., Waters, R.C., Strohecker, C., Schabes, Y., Freeman, W.T., Torrance, M.C., Golding, A.R., & Roth, M., (1994). Demonstration of an interactive multimedia environment. *Computer, 27*(12), 15–21.

Riel, M. (1994). Educational change in a technology-rich environment. *Journal of Research on Computing in Education, 26*, 452–474.

Ritchie, D., & Wiburg, K. (1994). Educational variables influencing technology integration. *Journal of Technology and Teacher Education, 2*, 143–153.

Sanders, M. (1994). Technological problem-solving activities as a means of instruction: The TSM integration program. *School Science and Mathematics, 94*(1), 36–43.

Snider, R.C. (1992). The machine in the classroom. *Phi Delta Kappan, 74*(4), 316–323. In *Assessing the role of technology in education*, by Wishnietsky, D. H. (Ed.) (1994). Bloomington, IN: Phi Delta Kappa.

Solomon, G. (1994). Math, science, and more. *Electronic Learning, 13*(7), 20–21.

Thornburg, D.D. (1992). *Edutrends 2010: Restructuring, technology, and the future of education*. San Carlos, CA: Starsong Publications.

West, P. (1994, December 7). Academy unveils draft of national science standards. *Education Week, 14*(14), 1, 9.

Whitaker, L. (1995, February). Aim straight at the curriculum. *Electronic School*, A41–A42. Supplement to *The American School Board Journal, 182*(2).

Wiburg, K. (1994). Relating teaching goals, student outcomes, and technology use. *Journal of Technology and Teacher Education, 2*, 227–235.

## Author Note

**Ronald E. Converse** is currently the director of science, mathematics, and health for Conroe Independent School District in Texas. He has been involved in science education for 30 years in four different school districts in three states as a teacher and K-12 science coordinator. He received the NSSA/Prentice Hall Outstanding Science Supervisor Award in 1990. He is heavily involved in efforts to reform science education at both the state and national levels.

# Science Using Networked Computers and the Internet

Jackie Shrago

Many of the reform efforts that abound in education give significant guidance as to how we attempt to respond to the need for change. The use of networked computers and the Internet (the network of networks connecting an estimated 30 million people around the world) offers great promise. But it is important that we remember to understand the purpose of incorporating technology or any other tool in making changes. Our purpose is not necessarily to use technology but to achieve student outcomes that demonstrate that students are competent in scientific thinking and can solve problems.

Four frequently identified elements of reform efforts have particular appropriateness as we consider the use of the Internet for improving science learning and teaching:

(1) responding to job market challenges,

(2) being accountable,

(3) increasing quality and productivity, and

(4) providing access and equity.

## RESPONDING TO JOB MARKET CHALLENGES

It is difficult to engage in any dialogue with business and industry without discussing the changes that need to occur in education in order to produce students who better meet the needs of the world competitive marketplace. Responding to this job market challenge has certain characteristics that can be well addressed by using the Internet for science student outcomes. One requirement is that students be able to solve real world problems. Can teachers have access to real world data for use by students? Can students have access to experts, in addition to their teachers, who can provide guidance in solving problems? Can students have access to the most up-to-date materials and information for solving problems in the real world? Can students learn team-oriented skills that prepare them for teams in future jobs? Can teachers and students organize the learning process so that real world problems are permitted to cross disciplines as they work to solve them?

Answers to all of these questions are readily addressed when we consider examples of teachers using the Internet for science. Various examples have emerged—from single question interactions to more elaborate projects that span a significant amount of learning time for students. Let's look at a few that have emerged in Tennessee's use of the Internet.

A teacher who was trying to stimulate his students to ask questions began to work with a local group of scientists who were interested in assisting local teachers. The problem, however, seemed that the time for scientists to leave their respective jobs and visit classrooms was very limited. By using an electronic bulletin board on a university network computer, the teacher could have any student pose a question. The teacher could then forward it electronically to one of the experts. The responses were detailed and focused on the age of the student asking the question. It was clear that students were receiving answers much more elaborate than would be possible across many scientific areas that the teacher alone could provide. In addition, since any trained teacher can have electronic access to this network connection, all of their students can have access to this mechanism. Training is free and available across the state from many public and private universities, as well as being offered by many K–12 teachers who have become trainers. In many situations, teachers found that their students asked better questions (i.e., beyond those whose answers can readily be found in a dictionary or encyclopedia) and are more intrigued with science because of the responses received from the "real" scientists.

This capability has proceeded over several years, with scientists participating from various universities and several federal research laboratories from around the country. They have willingly offered to respond to e-mail questions in their field of interest. The scientists were initially identified through a professional association in the Nashville area. The coordination is handled by a few participating teachers who regularly check the bulletin board and send the questions to an appropriate scientist who has agreed to participate. The answers then are posted for all students to see both questions and responses. It maximizes the participation of the scientists and incorporates experts into many more classrooms than was possible when the scientists started volunteering to participate. Some students found that questions and answers sparked their interest and generated a science project with more independent work. For all students, the sense that science emerges from questions rather than following the table of contents has generated more curiosity. Some of the questions originated from students watching television news and commercials.

Another more in-depth interaction occurs when a class can become electronically linked to a specialist and works over a period of time on a specific project. As an example, a university made "subject matter experts" available over a semester period. Studying the rainforest was of particular interest and fit especially with third grade objectives for one Tennessee teacher who became aware of the service. As the students organized themselves around components of studying the rainforest, they used a variety of resources to help them understand rainforest issues. In this situation, many of the documents they found on the Internet were too complex for them to read and understand, but the subject matter expert helped them interpret the materials while still letting them grapple with some of the differences of opinion in materials they have researched.

This kind of full-blown project involving research and access to experts increases the depth that students experience in studying a problem. It also helps students recognize that all answers are not simple and that often even experts disagree. Reconciling these disagreements encourages students to think—developing the kinds of skills that business and industry say are often left out of the education process. While this example emerged from the University of Texas, many more are possible and available. Subject matter experts need to be prepared to respond to questions on a regular basis and to respond using vocabulary that their linked class can comprehend. The teacher is responsible for managing the project in such a way that students work in cooperative groups and have a variety of resources available to them. Students only contact the subject matter expert when no other source is available. Students learn to work in teams and, of course, the problem is cross-disciplinary. Students, in this example, are developing scientific research skills, learning the impact of events in one part of the world on others, and incorporating social studies and reading and writing skills while understanding a real world problem.

Some of these examples are found through the Internet. Others use the bulletin board function that has emerged through community efforts in Tennessee to make a simpler utility available to K–12 teachers. Both have stressed content over technology, and student learning has evolved with teacher creativity and interaction over electronic means. No grant funds have been involved, and teachers have found each other through ordinary human networking as well as finding each other on the computer network.

## BEING ACCOUNTABLE

The discussion on evaluation and assessment of students is very complex and generates much disagreement among educators and other stakeholders. One element, however, is generally recognized as an important component of accountability: students publishing the results of their efforts as a measure of their performance and understanding of the learning process. Here the Internet has some unique advantages.

Tell a student to produce a paper and you have a requirement specified by the instructor. It requires self motivation to produce, and often some coercion by both school and home adults interacting with the student in order to obtain the desired outcome. But teachers continuously report that when the results of a project are going to be published on the Internet, an important factor is added: pride in the results, and perhaps some peer assessment that encourages students to produce their best work. Teachers consistently report that students are likely to produce multiple drafts before they are satisfied with the one that will be placed on the Internet. This phenomena seems to occur at any grade level of student, whether young or older. Perhaps we can see that all of us respond to wanting to "do a good job" if it is going to be viewed by others. This external assessment is motivation to produce, and it often delivers better results than a paper just for the teacher or even just for a grade.

The students who worked on the rainforest project wanted to produce their best work and have it viewed by their subject matter expert. The advent of multimedia capability for displaying visuals, sound, and textual information using the Internet's World Wide Web has dramatically changed students' ability to present to a much wider audience. Teachers and students with personal computers and a full connection to the Internet can produce multi-media reports easily. This publication opportunity provides an appropriate culmination to projects, encouraging students to synthesize in addition to researching.

Upon making the World Wide Web available to a limited number of teachers, several found that student research could be readily presented via this method. One high school class published its results on black hole research, where each team of students produced a document summarizing their respective team findings. Not only could they incorporate graphics for descriptive purposes, they included actual pictures taken from NASA satellites and made available on the Internet. They even linked to some NASA World Wide Web sites that continually update photos, so their reports are constantly updated with the latest scientific information. Their grasp that this area of inquiry is a constantly changing phenomena was a new insight for the students.

## INCREASE QUALITY & PRODUCTIVITY

More adults are becoming aware that the exponential pace of newly available scientific information makes it impossible for the average student to know enough to be an informed and educated adult without a commitment to life long learning. The biology studied by a typical ninth or tenth grader today includes significant emphasis and understanding of living cell biology that was, for the most part, unknown 20 years ago. How will it ever be possible for teachers, parents and textbook publishers to stay abreast of such a rapidly changing field? The Hubble telescope, within the last year, has completely changed explanations of how the universe began. Such vast and critical information makes traditionally published texts out-of-date for many areas of science.

Consider a parallel situation. The manufacturing sector of our economy has changed so dramatically in the last 20 years that most manufacturing processes require assembly of parts. Parts are manufactured by suppliers with specialty capabilities. Production, shipping, scheduling, and assembly are an elaborate process, performed by multiple companies and interwoven contracts with many inputs to achieve the end product usable by consumers. This process has become known as "just in time," indicating that no inventories are held. Rather, parts are ordered "just in time" to meet customer orders, dramatically reducing the cost of manufacturing and guaranteeing maximum productivity of people and inventories while meeting customer demand.

Applying the concept of "just in time" manufacturing to education provides some interesting parallels to the manufacturing process. Our knowledge base for many scientific areas of study is now dependent upon many resources. Certainly, a "textbook only" approach in any K–12 grade classroom will not produce the level of quality needed by today's graduates if they are to be knowledgeable productive citizens. But if we provide students with access to resources available on the Internet, the complex set of new information and ideas becomes available, often "just in time" for student learning. The Internet can be an overwhelming set of information; it is not particularly organized, and certainly not organized for a typical K–12 grade student. The teacher's job becomes one of coordinator of information, not provider of information, much like the manufacturer providing goods to the end consumer. It is also one of making certain that key concepts are understood so that current scientific endeavors, which often present conflicting information, can be placed in a context for student inquiry, research, and evaluation. It may often appear like "assembly," but certainly teacher adjustments for individual students and the nature of the content are a critical component to the end result of the learning process.

So much more information is available by using the Internet, but so much more is possible to understand. But perhaps, more importantly, the sense of "completing the textbook" is an obsolete notion. The Internet offers unlimited and continuous inquiry, debate abounds, new information shatters old ideas and hypotheses. Science is possible to experience first-hand as a life-long learning process. Information gathering skills, comparison skills, and synthesis skills become those that are practiced most, not memorizing facts that are subject to change. In a sense, we are "reengineering" the process of learning by encouraging access to the Internet. What students learn today will most likely be obsolete or at least enhanced and changed significantly within a few years. Learning is not complete when the book is done, or even the project. The learning process is most important. Producing a synthesis of findings and presenting those findings become critical skills, for the content is likely to change, but the searching, organizing, and evaluation steps remain. What better way to allow this process to emerge but to submerge students in the first hand sources of Internet information produced by scientists at research universities and federal laboratories?

The learning process in the Internet environment emphasizes active student involvement. The possibility for more in-depth understanding will often surpass a teacher's knowledge, especially if students are given the freedom to pursue topics of their own choosing. This is threatening for some teachers because we have spent so much time emphasizing the right answer. The teacher has been expected to have the right answer, if not in his or her head, certainly in the answer book provided by the publisher. With this strategy we have ignored the reality that when students are actively involved, they learn more because it is the *process* that leads to a richer and deeper comprehension, not which material you read or have presented to you.

## ACCESS AND EQUITY

Productivity in today's world also implies a need for current information. Even by the time this article is published the lists of possible World Wide Web sites and Listservs (mailing lists to which people can subscribe, organized around specific topics) will be incomplete or different. But their ever-changing nature reinforces the ability and opportunity to find current information through this medium. We are all aware of school libraries with books that anticipate man landing on the moon. The library books may be obsolete, but the library, if connected to the Internet, and the learning process are not.

One of the most significant opportunities presented by the Internet and its vast array of information and possibilities is the access that can be made available to all students. We certainly have some worries that all of the Internet will not be accessible because often rural and highly urban schools with the most significant learning challenges are also the most isolated from access to new technologies. Laboratories are the poorest in such areas. Teachers are often "burned out." Fewer resources from parents make enrichment trips and experiences in science museums less likely.

In Tennessee, we have demonstrated that every teacher and school librarian can have access to the Internet and that it is rather inexpensive. By connecting all teachers via dial-up access to the statewide university computer system, access is possible. Training is possible, often using volunteers, for any teacher to begin the learning curve in using the Internet. Networking among teachers as professionals helps to overcome some of the gaps. Rural students can be connected to peers and mentors in other grades and to subject matter experts. The opportunity for real science to be included in elementary classrooms is achievable, even where K–5 grade teachers have little personal experience in science. They can connect to bulletin boards, university experts, projects, and other classes to exchange ideas and stimulate each other's thinking.

Such access in Tennessee is provided through local dial-up modem pools in 13 cities and towns and for approximately 28 percent of the rural counties through an 800 number. Ultimately, more networking must be accomplished to provide the best types of Internet access, but immediate access and a beginning is achieved with this simple, relatively inexpensive technology. World Wide Web is not available on this network—yet. But as more teachers who are connected via the most basic dial-up capacity learn of its resources, more will assist school boards and decision-makers to obtain the resources for school-wide networking and World Wide Web access. Access to the Internet by individual schools can also be achieved through commercial services. Community and statewide groups can obtain information from U.S. Department of Commerce's Report on National Information Infrastructure. By following such a strategy of providing basic access and basic training, the opportunity emerges for teachers, students, and parents to become acquainted with the network and seek expansion in capabilities. Rather than providing a "top-down" solution, the need is driven by those closest to each school and each student. All are learning and becoming more familiar with the use of these technologies.

One major example of this strategy has emerged around the River Projects across Tennessee. Three years ago, one curriculum coordinator became aware of river exploration as a means to science instruction with hands-on activities and cross-curricular implications. Electronic access to information was also included. Using a bulletin board feature, some of the initial teachers were able to share their "river experiences." The idea spread. Another group of teachers followed the first example because they learned of the project through an electronic bulletin board. They got some training on using the river as a curricular focus and enhanced the bulletin board so that most rivershed areas in Tennessee were included. Teachers from grades 1 through 12 got involved, including many who had only limited science training themselves in their own college experience. A third group of teachers in rural counties received grant funding to use the network as part of their method for sharing information across many miles and connecting with experts at the university. Now "river reunions" are occurring. Teachers are con-ducting professional development for other teachers, and the three projects are finding ways of connecting to each other as professionals and with their classes sharing information. Students have participated in "river congress" events where they share their findings in a true scientific exchange of information. This process is growing and continues to attract more teachers. It is also helping to justify the expansion of network connections across many counties. There is definitely more to come from these efforts.

## SUMMARY

All of these examples are only a beginning. They are also not the complete answer to the challenges of improving science learning in our K–12 grade classrooms. But the technology of networked computers and the Internet provides many examples of practicing teachers overcoming their inherent fear of technology, overcoming barriers of limited resources, and finding learning strategies that involve students in successful scientific experiences. They lean on the resources of each community and not just the efforts of national bodies. They incorporate the knowledge of the scientific experts across our research universities and federal laboratories. They thrive on the creativity of teachers helping each other and their students to discover new ways to learn and the continuing dialogue of scientific inquiry. What could be a better outcome?

## EXAMPLE INTERNET RESOURCES

Joel Helton and Pam Burish, teachers in metropolitan Nashville schools, provided assistance in creating and annotating this list. To use these addresses, an individual needs access to the Internet with a personal computer and software, such as Lynx, Netscape, or Mosaic, that permits browsing the World Wide Web. Text information can be viewed if the user can access Lynx through his or her Internet provider. Pictures and other visual information can be viewed if the user has a full connection to the Internet and is able to use a visual browser with software such as Netscape or Mosaic. In either case, by following the directions within the software, the following addresses will take the user to the respective World Wide Web sites.

Addresses on the World Wide Web are referred to as URLs and begin "http." It is important to note that World Wide Web addresses are constantly changing. The time between writing and publishing this article may mean that the addresses have changed and some of these are inaccurate by publication. Exploration, however, is still an option, and many of the web searching sites will assist the new user in identifying some interesting places to begin.

## World Wide Web Science Exploration in High School

(mentioned in article)

http://www.tbr.state.tn.us/~rodriguezb/index.html

## Explorer Home Page

(mentioned in article)

http://www.tbr.state.tn.us/www/exhp.html

Explorers consists of nine teachers from middle Tennessee. Each teacher's class created a variety of interactive, cross-disciplinary experiences. These experiences extend across grade levels, elementary school, middle school, and high school. Each class developed and published a home page and other exciting documents for all ages. Be sure to follow the links to the adventure that was shared by all.

## Chemistry Teacher Resources

http://www.anachem.umu.se/eks/pointers.html

This is an attempt to present a comprehensive list of information pertinent to chemistry teaching on the Internet. You are therefore invited to submit pages for indexing, as well as to correct any errors you may find in the links.

## Dinosaur Information/Dinosaur Digs

http://128.174.47.80/

http://ucmpl

http://ucmpl.berkley.edu/dilophosaur/intro.html

## Encyclopedia Britanica

URL: http://www.eb.com/

This is a demonstration of their research web site—you have to subscribe to get any benefit from it.

## Frog Dissection

http://curry.edschool.virginia.edu:80/~insttech/frog/ http://george.lbl.gov/ITG.hm.pg.docs/dissect/info.html

University of Virginia has quicktime movies and online color images in an interactive frog dissection program. The second listing has images and interactive forms to fill out about the ongoing dissection. Model lessons for all grade levels (K–12) integrate Internet-accessible math, science and technology resources. Contact Virginia Space Grant Consortium at 2713-D Magruder Boulevard, Hampton, VA 23666

e-mail: vsgc@pen.k12.va.us

phone: 804-865-0726

The cost is $16 for IBM or MAC disk (specify). The accompanying Educator's Guide is a NASA videotape on using the Internet in the classroom.

## Hands-On Universe

http://hou.lbl.gov/houtitle.html

This is an education program sponsored by the National Science Foundation and the Department of Energy that helps high school students perform genuine astronomical research in their classrooms. Students from around the world can request observations from an automated 30" telescope, select and download images from an archive of over 1,500 images, and learn the math and science involved in professional astronomy through Hands-On Universe curriculum.

## Hotlist of K–12 Internet School Sites

http://toons.cc.ndsu.nodak.edu/~sackmann/k12/html

## The Hub

http://hub.terc.edu/

(e-mail) hub-mail-services@hub.terc.edu (In the body of your message, type "info".)

An Internet resource for science and math education that provides services to help you publish reports, curricula, projects in process, and requests for proposals.

## Implicate Beauty—Art By Brian Evans

http://www.vanderbilt.edu/VUCC/Misc/Art1/Beauty.html

This page, by Vanderbilt faculty member Brian Evans, is a showcase of computational art created by Brian. The general idea is that mathematics can be used to generate imagery capable of stirring one's emotions in one way or another. This is very high quality work.

## Jamie Zawinski List

http://home.mcom.com/people/jwz/netscape-bookmarks.html

Zawinski also works at Netscape. Very interesting list with lots of places to go that might be worth a look.

## Jefferson County Schools (Louisville, KY)

http://jefferson.k12.ky.us/

## Listservs

http://www.nas.nasa.gov/HPCC/K12/edures.html

On telnet—sunsite.unc.edu (Log in as lynx, then type g and then this URL.)

A Web site of listings of education-related listservs with reviews and roundup of electronic journals (many of them free).

## MVA Home Page

http://www.vuse.vanderbilt.edu/~ayersmv/mva_home.htm

This page by Mike Ayers is composed primarily of links to other places.

## University Web Pages

Carnegie-Mellon University http://www.cmu.edu/

Case Western Reserve University http://www.cwru.edu/

Dartmouth http://www.dartmouth.edu/

Duke University http://www.duke.edu/

Johns Hopkins University http://www.jhu.edu/

Massachusetts Institute of Technology http://web.mit.edu/

Northwestern University http://www.nwu.edu/

Washington University (St. Louis, MO) http://www.wust.edu/

Vanderbilt University http:/www.vanderbilt.edu/

Fisk University http://www.fisk.edu/

Belmont University http:/acklen.belmont.edu/

MTSU http:/www.mtsu.edu/

## NASA Spacelink

http://spacelink.msfc.nasa.gov/

Very impressive links to a lot of stuff NASA has.

## NASA Site for Downloading Graphics

http://stardust.jpl.nasa.gov/planets/welcome.html

## The Online Educator

http://www.cris.com/~felixg/OE/OEWELCOME.html

The editors of The Online Educator built a Web page that is really a gateway—a link to classroom treasures on the Internet. This is the place for teachers to start their Web activities each week, with visits to each of the hot-list sites featured. A new list appears each Monday. It is also a place where educators will find hints for using Internet information in their classrooms.

## Resources for Educators, Students, and Parents

http://www.mtjeff.com/~bodenst/page1.html

Its connections are vast and varied. For example, in clicking on "Sites for Educators" you will be pleased to discover links to a wealth of web sites. Included in the table of contents for the educators' sites are such well visited places as "Ask ERIC Home Page," "ERIC Clearinghouse on

Reading, English and Communication," "K–12 Resource Options," "NCSA Educational Resources Hotlist," "Vocal Point Online Student Newspaper," and "Steve's Dump - loads of educational resources."

### Science and Math

http://www.c3.lanl.gov:6060/SAMI-home

Want to visit the Field Museum of Natural History in Chicago?:

http://www.bvis.uic.edu/museum/

How about the Franklin Institute Science Museum in Philadelphia?:

http://sln.fi.edu/tfi/welcome.html

To improve middle school science, check out:

http://www.clp.berkeley.edu/CLP.html

### Social Studies

http://execpc.com/~dboals/boals.html

### Teachers Guide to the U.S. Department of Education

http://www.ed.gov/pubs/TeachersGuide/index.html

Includes all kinds of resources: Eisenhower Clearinghouse, National Diffusion Network, parent resources, etc.

### Unusual Finding

http://ghg.ecn.purdue.edu/

## Author Note

Jackie B. Shrago is vice chancellor for information technologies at the Tennessee Board of Regents where she develops plans for the infusion of technologies in higher education curricula and increased access to higher education opportunities for all Tennesseans. Previously, Ms. Shrago organized the technology transfer function and coordination of K–12 technology efforts at Vanderbilt University. She has also served as presidential campaign comptroller and state director for Senator Al Gore, and as co-founder of the Tel Research Corporation.

# Strategies for Implementing Computer Technology in the Science Classroom

Paul E. Adams
Gerald H. Krockover
James D. Lehman

All science education reform efforts directed at the improvement of K–12 science instruction in schools include technology components that move beyond the traditional drill and practice on an outmoded computer platform. Whether that reform effort is *Science for All Americans* from the American Association for the Advancement of Science, the National Science Teachers Association's *Science, Scope and Sequence*, or the National Research Council's *Science Standards*, one common thread is the need for technology infusion. The technology infusion suggested usually revolves around (a) access to information via the Internet or a commercial service such as Prodigy or CompuServe, (b) the use of technology for simulations and laboratory based experiences, and (c) the retrieval of media sources via CD-ROM or laser disc. This paper features examples of suitable technology for integration into the science classroom. To achieve this integration, all science classrooms as a minimum should have at least (a) one computer with adequate RAM and hard disk space, (b) a CD-ROM drive and laser disc player, (c) a high speed modem with a dedicated telephone line, (d) laboratory and simulation hardware/software, and (e) a display device for use with large groups. The science teacher

and students should also have the ability to use word processing, spreadsheets, organizers, graphics, and desktop printing programs.

The ways to use computer technology in the science classroom are many and varied. Taylor (1980) proposed a scheme for classifying computer uses in education into three broad categories: tutor, tool, and tutee. While distinctions among these categories are blurred in many cases today, this framework is still useful for discussion. Tutor applications, commonly referred to as computer assisted instruction, are those in which the computer instructs the student. Tool applications are those in which the teacher or students use the computer as an aid in performing typical work tasks such as writing, acquiring, and organizing data, or performing calculations. Tutee applications are those in which the student "teaches" the computer to do something, usually through the use of a programming language.

Table 1 identifies ways to use computers in the science classroom, organized according to Taylor's three categories and further divided into those applications that are applicable for large group, small group, and individual instruction. Naturally, a number of applications can fit into more than one cat-

66

National Science Teachers Association

egory. Videodiscs, for example, can be used to provide direct instruction to large groups, but they can also be used as a tool with large groups (e.g., for testing), and they may be involved in small group or individual multimedia instruction. Microcomputer-based laboratories (MBLs) are probably best used as tools for data gathering and presentation by small groups of students, although individual and large group methods of use are possible. Table 1 suggests ways to use computers and allied technologies in the science classroom, but it is not intended to be restrictive. Teachers and students may find other useful ways to employ computers. The remainder of this article will focus on some of the newer and less familiar of these applications for science teaching and learning.

## MICROCOMPUTER-BASED LABORATORIES

One of the most significant developments in the use of computers in the science classroom has been the advent of the microcomputer-based laboratory (MBL) (Tinker, 1987; Stringfield, 1994). MBLs use the microcomputer for data acquisition, display, and analysis. Data are collected through the use of probes or sensors that are connected to the microcomputer. Depending on the software available, the data can be displayed in tabular or graphical form, in real-time or delayed-time. The collected data can be analyzed graphically, statistically, or mathematically to investigate scientific phenomena if the appropriate software is available on the microcomputer.

Table 1. *Uses of Computers in the Science Classroom*

| LARGE GROUP | | |
|---|---|---|
| **Tutor** | **Tool** | **Tutee** |
| • Demonstrations | • Demonstrations | • Hypermedia project development |
| • Graphing software/ calculators | • Graphing software/calculators | • Spreadsheets or computer programs for problem-solving |
| • Presentation hardware/ software | • Presentation hardware/ software | |
| • Videodiscs | • Videodiscs | |
| **SMALL GROUP** | | |
| **Tutor** | **Tool** | **Tutee** |
| • Groupware (e.g. Tom Snyder software) | • Databases | • Hypermedia project development |
| • Multimedia (CD-ROM and videodiscs) | • MBLs | • Spreadsheets or computer programs for problem-solving |
| • Simulations | • Multimedia (CD-ROM and videodiscs) | |
| | • Spreadsheets | |
| | • Telecommunication | |
| **INDIVIDUAL** | | |
| **Tutor** | **Tool** | **Tutee** |
| • Drill and practice | • Databases | • Hypermedia project development |
| • Multimedia (CD-ROM and videodiscs) | • Graphing software/calculators | • Spreadsheets or computer programs for problem-solving |
| • Simulations | • Multimedia (CD-ROM and videodiscs) | |
| • Tutorials | • Spreadsheets | |
| | • Telecommunication | |
| | • Word processors | |

**Table 2.** *Commercially Available Sensors and Probes for MBL*

| | | | |
|---|---|---|---|
| • motion detector | • pH electrode | • radiation | • thermocouple |
| • microphone | • colorimeter | • strain gauge | • barometer |
| • light sensor | • magnetic field | • heart rate | • photogates |
| • voltage | • relative humidity | • temperature probe | • force probe |
| • smart pulley | | | |

The goal of using MBLs is to assist students in acquiring science skills and concepts by performing experiments in a manner similar to the way scientists perform actual research. This is accomplished through a redistribution of the time needed for laboratory (Amend, Tucker, & Larsen, 1990). MBLs reduce the time needed for tedious data acquisition and organization, thereby allowing more time to be devoted to experimental design and interpretation. The research literature indicates that, efficaciously used, MBLs are effective in helping students develop graphing skills, critical thinking skills, graph interpretation skills, acquisition of science concepts, and collaborative learning skills (Amend et al., 1990; Stuessy & Rowland, 1989; Tinker, 1987).

## A Typical MBL System

An MBL system typically consists of a microcomputer, probes, an interface box or card, and data acquisition and analysis software. An interface box may not be needed, since some systems interface through the game ports or communication ports. Commercial MBL interfaces and probes can be purchased for a specific brand of microcomputer or, if an individual has the time and expertise, the interface and probes can be constructed.

Microcomputer-based laboratory systems have been developed for most popular computer systems such as the Apple II series, Macintosh series, or the IBM series and its compatibles. Scientific equipment suppliers,

such as Pasco Scientific, Central Scientific, Sargent-Welch, TEL-Atomic, and Vernier Software, offer a full range of interface boxes or cards, probes, and acquisition and analysis software. The higher-end systems will generally allow the user to export the collected data file into word processing, spreadsheet, or graphing programs.

## Uses of MBL

Microcomputer-based laboratory systems can be used in any of the sciences. Over the last few years there have been several new probes made to help in the data acquisition of a wider range of physical, chemical, and biological phenomena. A listing of the types of probes that are currently available is provided in Table 2.

These probes, when used singly or in combination, provide a wide range of laboratory activities (Podany, 1990). For example, respiration rate can be determined by using a temperature probe placed just inside the nostril to measure the temperature of exhaled and inhaled air as a function of time in order to produce a temperature versus time graph (Stringfield, 1994). Many schools across the country measured temperature change as a function of ambient light during the solar eclipse in 1994 using light sensors and temperature probes. A motion detector can be used by students to act out the motion represented by various distance versus time or velocity versus time graphs. A force probe, when connected by a string to a toy airplane in a circular orbit, can be used to measure the tension in the string necessary to constrain the plane to its circular orbit. Using a colorimeter, students can investigate the concentration of a solution. Many of the commercial MBL packages provide laboratory ideas and/or manuals.

## New Developments in MBL

One new development for MBL has been the addition of digitized video. Students can videotape an event, such as the tossing of a ball or a bird taking flight, and by using a video digitizing board connected to the computer, produce a digitized video. This video can be analyzed through use of appropriate software. Although this application in the science classroom is relatively new, indications are that it can be a very productive means of learning science (Beichner, 1990).

Another development has been the advent of calculator-based laboratories (CBL™) by Texas Instruments. The CBL™ System is a hand-held device that uses MBL probes with appropriate adapters, permitting students to collect data in a variety of real-world settings. The collected data can be exported from the CBL™ system to a TI-82 or TI-85 graphing calculator for subsequent display and analysis. The data can also be exported to a IBM-compatible or Macintosh computer. Although CBL™ is still in its infancy, it might prove to be a cost-effective way of incorporating technology into the classroom, because each laboratory could be setup for about one-third the cost of MBL. However, CBL™ does not have the versatility and ease of use available in MBL systems.

## TELECOMMUNICATIONS

To utilize telecommunications, one must have access to a computer. That computer must have the ability to contact other computers via a telephone line or dedicated network connection. To use a telephone line, either an internal or external modem is needed. This modem serves as the "translator" allowing computers to interact with

each other. To implement telecommunications in schools, several strategies may be followed. An existing telephone line may be used, or a telephone line may be split into two lines by the telephone company, allowing the computer to have a dedicated telephone number that is not for voice calls. Enhanced services such as call waiting and call forwarding may need to be disabled, because these enhanced services tend to interfere with the ability of computers to communicate with each other.

Many telephone companies provide grants to schools to make classroom telephone lines available at low cost. However, while vendors promote the necessity of a telephone jack in proximity to the telecommunications hardware, telephone lines may actually be run great distances from the telephone jack without any loss of data reception or transmission capability simply using telephone line and plug-in connectors. If it is not feasible to place a telephone line in the science teacher's classroom, good locations to tie into telephone lines may be areas such as the school media center, the athletic department office (since most coaches are on-line to schedule athletic events), the principal's office, the school lunchroom office, or the school district business office.

The computer user must also have software to conduct the telecommunication activity desired. Software may be generic to allow access to a variety of services. Examples of generic types of telecommunication software are Procomm, Smartcom, Z-Term, or Crosstalk. Specific software provides access to one particular service. Examples of specific services are America Online, CompuServe, Genie, Delphi, and Prodigy.

Modems transmit information at a variety of rates. In general, the faster the modem transmits information the greater the cost of the modem. However, higher speed modems are more desirable because information can be obtained much faster, and this controls cost when using services that assess charges. Most services now support 9,600 or 14,400 baud modems. Information acquired can be directed either to a hard drive, a floppy disk, a computer monitor, or a printer.

All science classrooms require access to the information highway called the Internet. The Internet is a collection of access points for information acquisition. Whereas access to the Internet was previously limited to higher education and the business community, it is quickly being made available to all who are interested in participating. Furthermore, the Internet is moving toward offering a more user-friendly environment in which to operate. The Internet provides access to worldwide electronic mail, libraries and governmental agencies, databases, bulletin boards, and list servers. Furthermore, tools are being developed that allow individuals to successfully find information.

Tools such as Gopher and its search engine Veronica (see Figure 1 for a sample Gopher screen); file transfer protocol (ftp) and the Archie search engine of its sites; Wide Area Information Services (WAIS); and the World Wide Web (WWW) with appropriate software, such as Mosaic or Netscape, allow users to access information in an organized manner.

## MULTIMEDIA AND HYPERMEDIA

Among the most exciting of the recent developments for science teaching and learning is the proliferation of multimedia and hypermedia. Today, multimedia refers to any system that makes use of multiple media— text, graphics, audio, and video. Hypermedia, made widely popular after the development of Apple's HyperCard program, is a special class of multimedia that relies on information links activated through the click of a mouse on a software button. Products incorporating these ideas are widely available, and

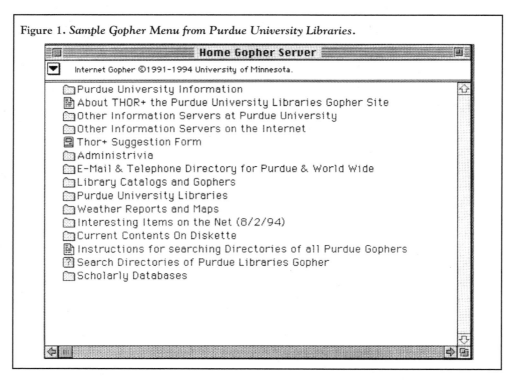

Figure 1. *Sample Gopher Menu from Purdue University Libraries.*

the hypermedia concept has come to the Internet in the form of Mosaic, Netscape, and the World Wide Web.

The current multimedia explosion began with the laser videodisc, a medium that remains very useful in the science classroom. A standard 12-inch videodisc can hold up to 30 minutes of motion video and 54,000 still images, as well as audio. Any frame on a videodisc can be instantly accessed, and videodiscs do not wear out. Videodiscs are an excellent medium for group use in classrooms, and they can be used by small groups and individuals as well. Packages such as Optical Data's *Windows on Science* series, Videodiscovery's *Bio Sci II*, National Geographic's *STV: Restless Earth*, and Tom Snyder's *Great Solar System Rescue* offer a wealth of images that are useful in science instruction.

The newer cousin of the videodisc, the CD-ROM, is taking the computer world by storm. A CD-ROM is a compact disc that acts as a mass storage device for a computer. One CD-ROM can hold about 600 megabytes of computer information—enough to store an entire set of encyclopedias. The information can include text, graphics, audio, and even limited video. Thousands of CD-ROMs are now available for both Macintosh and IBM and compatible computer platforms. In science, many CD-ROMs are available for both reference and instruction, including Maxis' *Redshift* astronomy, Software Toolworks' *The Animals* (from the San Diego Zoo), Tom Synder's *Hip Physics*, and Time Warner's *LIFEmap* series.

In addition to making use of existing multimedia products, students can become involved in developing their own multimedia reports and projects through the use of hypermedia authoring tools. Popular ones include Apple's HyperCard and Roger Wagner's HyperStudio for the Apple Macintosh, IBM's Linkway and Linkway

Live for DOS, and Asymetrix Toolbook for Windows. Using these products, students can create multimedia reports and projects that include text, graphics, audio, and even video from a laser disc or in digital form. In this process, students become actively involved in understanding, organizing, and presenting scientific information.

## PRESENTATION HARDWARE AND SOFTWARE

Surveys of the use of computers in science teaching and learning (Baird, 1989; Becker, 1991) suggest that many science teachers must deal with a paucity of computer resources. A relatively recent development that supports the use of a single computer with an entire classroom of students is presentation hardware and software. While normally a computer display is intended for viewing by only a single individual, special equipment is available to permit the computer output to be displayed in a way suitable for viewing by groups. Options include use of a large computer monitor, large television or video monitor (usually requiring special computer-to-video conversion hardware), video projector, or liquid crystal display (LCD) panel. Each of these options has advantages and disadvantages. For schools, LCD panels are an attractive option because they are reasonably priced (starting under $1,000) and can be used with standard overhead projectors. Alternatively, some computer-to-video conversion devices cost only a few hundred dollars and can be used to display computer output on a standard video monitor. The only drawback to this technique is that some loss of resolution usually occurs, and it may be necessary to use larger fonts on the computer for optimum viewing on a large video monitor.

Once a school has acquired presentation hardware, it can be used with small or large groups of students for group use of instructional software, demonstrations, presentation of student projects, and the like. In fact, any of the applications discussed to this point could be used with presentation hardware. In addition, a special class of software, called presentation software, has been developed specifically to take advantage of the computer's capabilities to act as a group presentation tool. These programs permit the creation and display of computer-generated frames of information, analogous to slides. Popular presentation packages include Microsoft Powerpoint, Adobe Persuasion, Lotus Freelance Graphics, IBM Storyboard Live, Asymetrix Compel, Wordperfect Presentations, and Harvard Graphics.

## CONCLUSION

There are numerous strategies science teachers can use to implement computer technology in their classrooms. To do this successfully will take the support of school administrators, local boards of education, and the community in order to overcome the traditional obstacles to change. Using new technologies requires news ways of thinking about teaching and learning. It also requires restructuring education from a textbook-driven curriculum to one driven by a desire to enhance the learning process.

## References

Amend, J.R., Tucker, K.A., & Larsen, R. (1990). Drawing relationships from experimental data: Computers can change the way we teach science. *Journal of Computers in Mathematics and Science Teaching, 10,* 101–111.

Baird, W.E. (1989). Status of use: Microcomputers and science teaching. *Journal of Computers in Mathematics and Science Teaching, 8*(4), 14–25.

Becker, H.J. (1991). Mathematics and science uses of computers in American schools, 1989. *Journal of Computers in Mathematics and Science Teaching, 10*(4), 19–25.

Beichner, R.J. (1990, April). *The effect of simultaneous motion presentation and graph generation in a kinematics lab.* Paper presented at the meeting of the National Association for Research in Science Teaching, Atlanta, GA. (ERIC Document Reproduction Service No. ED 319 597)

Podany, Z. (1990, November). *Ideas for integrating the microcomputer with high school science.* Portland, OR: Northwest Regional Educational Laboratory. (ERIC Document Reproduction Service No. ED 328 443)

Stringfield, J.K. (1994). Using commercially available microcomputer-based labs in the biology classroom. *The American Biology Teacher, 56,* 106–108.

Stuessy, C.L., & Rowland, P.M. (1989). Advantages of micro-based labs: Electronic data acquisition, computerized graphing, or both? *Journal of Computers in Mathematics and Science Teaching, 8*(3), 18–21.

Taylor, R.P. (Ed.). (1980). *The computer in the school: Tutor, tool, tutee.* New York: Teachers College Press.

Tinker, R. (Ed.). (1987). *A preliminary sampling from the MBL source book.* Cambridge, MA: Technical Education Research Centers.

## Author Note

**Paul E. Adams** is a graduate student in science education at Purdue University. He has utilized microcomputer-based laboratories and video technology extensively in the instruction of physics.

**Gerald H. Krockover** is a professor of earth and atmospheric science education at Purdue University. He has integrated technology into his pre- and in-service teacher programs.

**James D. Lehman** is an associate professor of educational computing at Purdue University. He has implemented technology education into both K–12 and higher education programs.

# Action Research in the Science Classroom: Curriculum Improvement and Teacher Professional Development

### Donna F. Berlin

Among the significant challenges identified by recent education documents is the lack of communication between researchers and practitioners—bridging the gap between research and practice. White and Tisher (1986) reflect that while a great deal of science education research has been conducted over the last decade, very little has informed or affected practice. To meet this challenge, they suggested that teachers become full members of research teams. "This development may lead to a different, collaborative style wherein research is done by and with, rather than on, the teacher." (White & Tisher, 1986, p. 897) This type of collaborative research characterizes action research. Although action research has been defined in many different ways, the term as used in this chapter denotes systematic and recursive inquiry and reflection in a collaborative learning community directed toward the understanding and improvement of practice. Readers interested in a review of the action research literature are referred to Hollingsworth (1992), Holly (1991), Kemmis and McTaggart (1988a, 1988b), McCutcheon and Jung (1990), Noffke (1989, 1990), and Watt and Watt (1993).

Within the current science education reform environment, action research is being endorsed as a means to broaden the research base, expand knowledge, and strengthen the impact of research. Shymansky and Kyle (1991) provide a strong rationale for the use of collaborative action research in science education research.

"Our understanding of science teaching and learning will be enhanced by practitioners and researchers theorizing, planning, conducting, and interpreting research that is pedagogically valid. Enhanced communication and collaboration should inform the process and influence practice." (p. 40)

Klapper, Berlin, and White (1994) strongly argue that systemic reform must be linked to a professional environment that

... provides teachers the resources to continue extending (through both self-learning and external presentation) their mastery of content and pedagogy ... supports teachers engaging in critical self-reflection and analytic/systematic inquiry; and encourages teachers to pursue innovation within the classroom, school, school district, and the enterprise of teaching. (p. 3)

In a recent conference held at The National Center for Science Teaching and Learning (White & Klapper, 1994), one of the five organizing theme questions was "What are the appropriate relationships between practice and research in science education?" Both university researchers and practitioners agreed that

> for change in practice to occur, there needs to be collaboration between researchers and practitioners. For practitioners to use and value research, they must be a part of the process itself, actively contributing to the research enterprise. For research to be valuable to practice, real problems emerging from practice should become a part of the science education research agenda. (Berlin & Krajcik, 1994, p.1 )

Clearly, the National Science Teachers Association Board of Directors has made their commitment to the role of science teachers in action research explicit in the following four recommendations:

1. research should be a collaborative endeavor,

2. teachers should be action-researchers,

3. research must be close to the classroom, and

4. an investigative society should be created.

(Kyle, Linn, Bitner, Mitchener, & Perry, 1991, pp. 416–418)

Action research promises not only to improve practice, but also to contribute to the professional development of teachers and the professionalization of teaching. A national focus upon teacher professional development is articulated in the recent *Draft Mission Statement and Principles of Professional Development* (U.S. Department of Education, Professional Development Team, 1994). This document recognizes the essential role of practitioners in education reform and the need for enabling conditions, an environment that promotes and supports the professionalization of teachers. Among the ten professional development principles put forth, the following four are especially relevant:

1. respect and nurture the intellectual capacity of teachers and others in the school community;

2. enhance leadership capacity among teachers, principals, and others;

3. require ample time and other resources that enable educators to develop their individual capacities [*sic*], and to learn and work together; and

4. promote commitment to continuous inquiry and improvement embedded in the daily life of schools (U. S. Department of Education, Professional Development Team, 1994).

The *National Science Education Standards* (National Research Council, 1996) proposes standards for the professional development of science teachers. Support for teachers as researchers is evident in the standards and suggested learning experiences for teachers of science. Professional Development Standard B—Learning to Teach Science suggests that "teachers use inquiry, reflection, interpretation of research papers, modeling, and guided practice to build understanding and skill in science teaching" (p. III–12–13). Professional Development Standard C—Learning to Learn suggests that "opportunities are provided to know and have access to existing research and experiential knowledge" and that "opportunities are provided to learn and use the skills of research to generate new knowledge about science and the teaching and learning of science" (p. III–18).

Of special relevance is the discussion related to Standard C.

Teacher[s] develop habits of conducting formal and informal classroom-based research to improve their practice. They ask questions about how students learn science, try new approaches to teaching, and evaluate the results in student achievement from these approaches (p. III-21).

## BERLIN-WHITE ACTION RESEARCH MODEL (BWARM)

A model for action research, the Berlin-White Action Research Model (BWARM), was developed and implemented for six years (1987–1993) under the auspices of a State of Ohio Academic Challenge grant. The National Center for Science Teaching and Learning (NCSTL), funded by the U.S. Department of Education, Office of Educational Research and Improvement, enabled subsequent implementations in Grove City, Ohio (1992–1993, in collaboration with South-Western City Schools); San Francisco, California (1993–1994, in collaboration with the Far West Laboratory for Educational Research and Development); and Elkhart, Indiana (1994–1995, in collaboration with Goshen College and the Elkhart Community Schools). In addition, NCSTL supported a five-year longitudinal study of the BWARM Program to determine the attitudes and perceptions related to educational research and educational innovations and document the professional growth and development of participating teachers.

## Goal and Objectives

The primary focus of the year-long BWARM Program is action research designed to prepare and support teachers in the development, implementation, and evaluation of innovation within their classroom. In addition to curriculum improvement, the program seeks to "improve the structures and social conditions of practice" from a "professional critique" standpoint (Hollingsworth, 1992). The specific objectives of the program are to

1. provide teachers with knowledge and experiences related to innovative teaching methods and materials;

2. provide teachers with knowledge and experiences in order to conduct action research;

3. develop positive attitudes and realistic perceptions related to educational research and innovations in teaching and learning; and

4. develop, implement, evaluate, and disseminate innovative teaching methods and materials.

## Program Structure

The BWARM Program consists of three interrelated phases over a period of four academic quarters. These phases are pedagogical awareness; research, development, and evaluation; and classroom applications.

*Pedagogical Awareness.* This phase consists of two special topics courses offered during the summer quarter. The courses are designed to provide knowledge and experiences to advance teacher content and pedagogy learning and to serve as a springboard for the development of educational innovations. Science-related special topics have included alternative learning environments; science, technology, and society; and science education reform: implications for the K-8 classroom. Special topics can be selected from issues and initiatives relevant to the local educational community or state and national concerns.

*Research, Development, and Evaluation (R D & E)*. This phase consists of a third course offered in the summer quarter. This course, Action Research: Solving Educational Problems in the Classroom, prepares the teachers in the fundamentals of inquiry in education. It includes literature search strategies and basic concepts and principles of research design, data collection, data analysis, and data interpretation related to both quantitative and qualitative paradigms. It should be noted that this program, distinct from other action research programs reported in the literature (McCutcheon & Jung, 1990; Richardson, 1994; Watt & Watt, 1993) exposes teachers to a variety of research methods, both quantitative and qualitative. The underlying assumption is that science education research should include "... the full range of investigative methods, embracing quantitative research and qualitative/ethnographic/naturalistic research to address either basic or applied questions" (Kyle et al., 1991, p. 414).

*Classroom Applications*. During the academic year, three quarter-long seminars are provided to facilitate and support the transformation of the previous two phases. Biweekly seminars are designed to provide ongoing review and support for the teacher-researchers and continuous feedback for program modification. Autumn quarter focuses on development of curriculum innovations and data collection procedures, Winter quarter on classroom implementation and data collection, and Spring quarter on data analysis, interpretation, and report writing.

A culminating two-day conference brings together the teachers and other professional educators to share the curriculum innovations and action research results. These efforts are disseminated as two-part conference proceedings: a description of the innovation and a report of the research results related to the evaluation of the innovation. Some examples of the science action research projects are Using Student-Scientist Interactions to Improve Attitudes Toward Science and Science-Related Careers; The Effects of Personalization on Learning of Science and Mathematics; The Effects of Science Fairs in the Elementary School; Ways to Minimize Time-of-Day as a Factor in Student Performance [in Science Class]; Exploring Student Understanding of and Attitudes Toward the Environment through Real-World Activities; and The Effects of Using Graphing Calculators on Student Science Attitudes.

## Longitudinal Study

The teacher-researchers ($N$=92) in the five-year longitudinal study were primarily kindergarten through eighth grade teachers from eight counties in the state of Ohio. Semantic differential instruments were created, tested, and revised to measure change in teacher attitudes and perceptions related to educational research and educational innovations. Additional data sources include biographical and demographic information, responses to open-ended questionnaires, and taped teacher presentations/discussions of projects. Follow-up questionnaires, distributed approximately six months after completion of the program, and periodic follow-up communications were also analyzed to determine current attitudes and perceptions related to educational research and educational innovations as well as to gather indicators of professional growth.

Eighty-seven teacher-researchers completed action research projects. Based upon both quantitative and qualitative data, the results suggest that this year-long action research program (a) enhances teacher attitudes toward educational research; (b) promotes realistic perceptions related to educational research; (c) fosters positive dispositions toward educational innovations; (d) increases teacher involvement in local, state, and national professional activities as presenters and leaders; (e) increases grant-writing efforts and successes; (f) facilitates the implementation of educational innovations and improved teaching and learning in individual classrooms; (g) changes the participating teachers' views of their classroom roles to include reflection and inquiry; and (h) stimulates academic collaborations within school buildings, across school districts, and with university and business partners. (See Berlin & White, 1993, for a comprehensive report of the longitudinal study.)

## SALIENT FACTORS

Based upon the implementation and evaluation of the BWARM Program, four salient factors have emerged: collaboration, communication, support, and recognition. Collaboration is one of the most important aspects of action research. The BWARM Program assumes a community of learners of university faculty and practitioners. The university faculty serve as facilitators bringing their expertise, experience, and resources to the process. The practitioners are the driving force as they conceptualize, develop, implement, and evaluate their classroom innovations. One of the most cogent aspects of collaborative action research is the sharing of multiple perspectives and expertise.

Communication is an important aspect of collaboration that is often overlooked. Teachers and researchers have different personal theories of practice and research that are realized in different environments. The solution to "different speech communities" (Florio-Ruane & Dohanich, 1984) is not to simply translate for teachers, which perpetuates a one-way research relationship, but to enable the two groups to continually and openly communicate to build a consensual dialogue about the research.

Support for action research demands both time and resources (material and human). The report of the National Education Commission on Time and Learning (1994) recommends that "teachers be provided with the professional time and opportunities they need to do their jobs" (p. 36). This job includes both a school day and an academic day. The academic day provides opportunities for professional growth and development including time for all aspects of the action research process. To provide this time, the report suggests "... extending the contract year to pay teachers for professional development, using the longer day for the same purpose, or providing for the widespread and systematic use of a cadre of well-prepared, full-time, substitute teachers" (p. 36). Other strategies include "redefine the workload; promote time-sharing and teaming; provide flexibility in scheduling within the school day, week, term, and year; and restructure the school year to include a research quarter" (Berlin & Krajcik, 1994, p. 4).

Action research projects, particularly those designed to implement innovations in the classroom, require material resources. Two projects at NCSTL (the BWARM Program and the Science Teaching Partnership Project) affirm the need to provide resources including funds for practitioners to use at their discretion. These can include instructional materials; research equipment (e.g., video cameras, tape recorders, commercial instruments); and access to telecommunication (e.g., e-mail, electronic databases). Finally, human resources are needed to prepare teachers to design and engage in action research as well as provide critical reflection and analysis. Professional development opportunities such as action research courses, seminars, institutes, and forums for practitioners and researchers to jointly present and discuss their work are also suggested. Other options include graduate student teaching, graduate student research, and study leave of absence support for teachers.

The final key factor is recognition. Recognition can assume two forms: commendation and compensation. Both can provide personal and professional motivation and incentive to engage in action research. Specific strategies include

1. recognition of teacher-researchers through commendations by principals, superintendents, supervisors, school boards, etc.;

2. letters recognizing teacher-researchers to principals, superintendents, supervisors, and school boards;

3. public awareness and commendation of teacher-researchers and their projects through newspaper, video, and television coverage;

4. public acknowledgment and support of the "academic day" of teachers and their role as researchers; and

5. honorariums, course credit, and release time to engage in activities that extend beyond the school day.

These factors, though essential, are not commonly found in the majority of schools. A starting point to build an environment supportive of action research is to assess the current teaching environment. The Professional Environment for Teaching Survey (PETS; Berlin, Klapper, & White, 1994) is available from The National Center for Science Teaching and Learning, 1929 Kenny Road, Columbus, OH 43210. It can be used to develop a profile of the teaching environment in terms of opportunity for collaboration; facilitation of communication; availability of time and human, material, and fiscal resources; and provision for the recognition of professional teachers and their initiatives.

After ensuring the necessary support structure, publications such as those by Altrichter, Posch, and Somekh (1993); Calhoun (1994); Ross (1984); and Sagor (1992), and one in preparation based on materials used in the BWARM Program, can guide the action research process. Other sources of information include the work of the National Science Teachers Association Committee on Research in Science Teaching, Task Force on Developing a Research Agenda, and Task Force on Developing Science Education Research Databases for Teachers. Numerous papers related to action research are presented at both the National Association for Research in Science Teaching and American Educational Research Association annual conferences. Of recent vintage is the *Educational Action Research*, an international journal that publishes accounts of a range of action research and related studies, reviews of the action research literature, and dialogue on methodological and epistemological issues.

## CONCLUSION

Action research is integral to both the improvement of educational practice and the professionalization of teaching. As Watt and Watt (1993) state, "Teacher research as beneficial to the teacher involved, the children learning, the curriculum, and the broader social community no longer needs defense" (p.38). Now is the time to develop and implement action research programs for both preservice and inservice teachers and establish a teaching environment that provides the enabling conditions that nurture and support teacher-researchers engaged in action research.

## References

Altrichter, H., Posch, P., & Somekh, B. (1993). *Teachers investigate their work: An introduction to the methods of action research*. London: Routledge.

Berlin, D.F., Klapper, M.H., & White, A.L. (1994). Professional environment for teaching survey (PETS). Columbus, OH: National Center for Science Teaching and Learning.

Berlin, D.F., & Krajcik, J. (Eds.). (1994, March). *What are appropriate relationships between practice and research in science education?* Science Education Research Agenda (SERAC) Working Conference of The National Center for Science Teaching and Learning, Columbus, OH.

Berlin, D.F., & White, A.L. (1993, October). *Research-practice-policy*. Paper presented at the annual meeting of the International Conference on the Public Understanding of Science and Technology, Chicago, IL.

Calhoun, E.F. (1994). *How to use action research in the self-renewing school*. Alexandria, VA: Association for Supervision and Curriculum Development.

Florio-Ruane, S., & Dohanich, J.B. (1984, November). Research currents: Communicating findings by teacher/researcher deliberation. *Language Arts, 61*(7), 712–716.

Hollingsworth, S. (1992, August). *Teachers as researchers: A review of the literature*. East Lansing: Michigan State University, College of Education, Institute for Research on Teaching.

Holly, P. (1991). Action research: The missing link in the creation of schools as centers of inquiry. *Staff development for education in the '90's* by A. Lieberman & L. Miller (Eds.), (pp. 133–157). New York: Teachers College Press.

Kemmis, S., & McTaggart, R. (1988a). *The action research planner* (3rd ed.). Geelong, Victoria, Australia: Deakin University Press.

Kemmis, S., & McTaggart, R. (Eds.). (1988b). *The action research reader* (3rd ed.). Geelong, Victoria, Australia: Deakin University Press.

Klapper, M.H., Berlin, D.F., & White, A.L. (1994, Autumn). Professional development: Starting point for systemic reform. *COGNOSOS, 3*(4), 1–5.

Kyle, W.C., Jr., Linn, M.C., Bitner, B.L., Mitchener, C.P., & Perry, B. (1991). The role of research in science teaching: An NSTA theme paper. *Science Education, 75*(4), 413–418.

McCutcheon, G., & Jung, B. (1990, Summer). Alternative perspectives on action research.

*Theory into Practice, 29*(3), 144–151.

National Education Commission on Time and Learning. (1994, April). *Prisoners of time*. Washington, DC: U.S. Government Printing Office. (ERIC Reproduction Document Service No. ED 366 115)

National Research Council (1996). *National science education standards*. Washington, DC: National Academy Press.

Noffke, S.E. (1989, March). *The social context of action research: Comparative and historical analysis*. Paper presented at the annual meeting of the American Educational Research Association, San Francisco, CA. (ERIC Document Reproduction Service No. ED 308 756)

Noffke, S.E. (1990, April). *Action research and the work of teachers*. Paper presented at the annual meeting of the American Educational Research Association, Boston, MA. (ERIC Document Reproduction Service No. ED 320 871)

Richardson, V. (1994). Conducting research on practice. *Educational Researcher, 23*(5), 5–9

Ross, D.D. (1984, Winter). A practical model for conducting action research in public school settings. *Contemporary Education, 55*(2), 113–117.

Sagor, R. (1992). *How to conduct collaborative action research*. Alexandria, VA: Association for Supervision and Curriculum Development.

Shymansky, J.A., & Kyle, W.C., Jr. (1991, March). *Establishing a research agenda: The critical issues of science curriculum reform*. Report of a conference held April 8, 1990. Sponsored by the National Association for Research in Science Teaching and funded by the National Science Foundation. Atlanta, Ga.

U.S. Department of Education, Professional Development Team. (1994, October 31). *Draft mission statement and principles of professional development*. Washington, DC: Author.

Watt, M.L., & Watt, D.L. (1993). Teacher research, action research: The Logo action research collaborative. *Educational Action Research, 1*(1), 35–63.

White, A.L., & Klapper, M. (1994, Summer). An agenda for science education research. *COGNOSOS, 3*(3), 1–5.

White, R.T., & Tisher, R.P. (1986). Research on natural sciences. In *Handbook of Research on Teaching*, 3rd ed., by M.C. Wittrock (Ed.), pp. 874–905. New York: Macmillan.

## Author Note

**Donna F. Berlin** is an associate professor of elementary education at The Ohio State University. She is a research coordinator at The National Center for Science Teaching and Learning and the Mathematics Education Associate for the Eisenhower National Clearinghouse for Mathematics and Science Education. She was the principal investigator of a six-year action research project.

Writing of this paper was supported in part by The National Center for Science Teaching and Learning under grant #R117Q00062 from the Office of Educational Research and Improvement, U.S. Department of Education. Any opinions, findings, or recommendations expressed in this publication are those of the author and do not necessarily reflect the views of the sponsoring agency.

# Action Research: Creating a Context for Science Teaching and Learning

William R. Veal
Deborah J. Tippins

In recent years, many teacher educators have come to view action research as a viable approach to science education reform and improvement (Dana, 1995; Grundy, 1987; Kincheloe, 1991; Tobin, Tippins, & Hook, 1994; Vitale & Romance, 1994). In the ever changing field of science, teachers are always incorporating new ideas, curricula, and technologies into their classrooms. When teachers decide to make changes, they are actively engaging in "action research."

Educational research has typically been a university-initiated and controlled endeavor in which most ideas and outcomes of research were outside the classroom. The initial emergence of action research paradigms in the 1940s and 1950s represented a methodology that starkly contrasted with traditional university-based experimental research paradigms in which teachers were the "objects" of research (Chien, Cook, & Harding, 1948; Corey, 1953). The early action research focus was on the involvement of teachers as researchers who systematically engaged in inquiry stemming from their classroom-based dilemmas and issues. Although action research declined for a time beginning in the late 1950s, a renewed interest in action research (Lieberman, 1986; Oja & Smullyan,

1989) has rekindled its use in both preservice and inservice science teacher education programs. When teachers—as researchers—examine alternative teaching methods and approaches, they are empowered to create knowledge that is relevant to their professional lives and their classrooms.

As science teachers deal with ever-changing science, they must be able to change their classroom practices in order to keep pace with new technologies and content. For example, the science standards set forth by the National Science Teachers Association (1992) lean toward more interdisciplinary science instruction, a new direction for many. Involving teachers in action research is one of the best ways to assist them in keeping abreast of the developments (such as interdisciplinary teaching) reflected in current reform documents.

Sometimes science teachers must reflect on and alter their practices to incorporate current reform ideas. Consider, for example, the following scenario: A community is dealing with the prospects of accepting low-grade radioactive waste in a new dump and storage facility. The students' families and community will be directly affected by this political, financial, and societal issue. The science teacher, in collaboration with the social studies teacher, may decide to plan and implement a unit designed to facilitate student exploration of the potential impact associated with the development of the facility. After reflection and discussion, the two teachers decide that the best possible way to introduce all the science associated with radiation is to use an interdisciplinary approach emphasizing the physics and chemistry of radiation, the ecological and biological effects of radiation, the industrial processes involved with radioactive waste management, the effect of weather on the transfer and storage of waste, and the social, political, and economical dimensions of waste management. In order to implement this unit, the teachers would have to change their traditional instructional approaches and curricula. Documentation of this practitioner-initiated change would provide valuable knowledge of the needs and challenges of this process for other educators.

The purpose of action research, as illustrated in the above example, is to provide teachers and schools with strategies that will facilitate new ways of teaching and thinking about science teaching and learning, so that ultimately learning can be enhanced in a critical, constructivist manner. Action research challenges the dominant research norm that excludes practical and emic (insider's perspective) knowledge. The nature of professional knowledge suggests that teachers, together with their students, should be the ones who conduct the research, collect, frame, and analyze data, and implement changes at classroom and school levels. As our previous example suggests, action research gives both teachers and students a voice in the content and curriculum; as the two teachers shift to a more project-centered approach to teaching, students are able to select which areas of the radioactive waste management unit they would like to study. From a constructivist perspective, students will be able to build on their prior knowledge and interests to study a subject that is personally relevant. In this sense, action research is consistent with the visions of professional teaching environments found in the major science education reform documents—one in which both teachers and students assume a decision-making and leadership role with regard to learning.

Action research is not only a different method, but is epistemologically an alternative viewpoint. Through action research, teachers and their students are empowered to construct personal epistemologies (knowledge bases) of the nature of science, teaching, and learning. Science teachers are fortunate to be in a field where inquiry, experimentation, and change are not foreign concepts.

## ACTION RESEARCH: DIVERSE PERSPECTIVES

Action research can be defined in many different ways; some of the most frequently encountered definitions include *collaborative inquiry* (Sirotnik, 1988), *teacher research* (Lytle & Cochran-Smith, 1992), *reflection* (Schon, 1987) and *reflective problem-solving* (McKernan, 1987), and *practical inquiry* (Richardson, 1994). In all of these definitions, the science teacher is a participant in research rather than an object of research. A general definition is that "action research is an ongoing, self-reflective process that involves a critical examination of teaching practices or theories to improve both personal practice and the education of students" (Hamilton, 1995, p. 79).

*Collaborative inquiry* usually involves the cooperation of the teacher and a university professor, but can include collaboration among teachers, schools, and districts. Research is done in the classroom and is shared with other participants.

*Teacher research* is a systematic, intentional inquiry by teachers about their own practice and work (Lytle & Cochran-Smith, 1992). Calhoun (1991) describes three types of teacher research: individual, collaborative, and schoolwide, all of which originate with the science teacher.

Dewey (1993) was the first to describe *reflection* in teaching as a specialized from of thinking that results from a direct experience and leads to purposeful inquiry and problem resolution. Schon (1987) elaborated upon this idea with his explication of reflection-in-action. Reflection-in-action describes the teacher's actions as generated and tested on the spot, while reflection-on-action involves post-hoc thinking and deliberation.

Action research, when defined as *practical inquiry*, is undertaken by a science teacher who changes a classroom practice in order to enhance the social atmosphere for learning. Most science teachers are adept at research practices as they relate to science experiments; they must learn to apply this inquisitive approach in making changes to their practice. Practical inquiry can bring about quick and meaningful change, because the change is both generated and implemented by the teacher. In the process the teacher may develop a sense of empowerment that can serve as a basis for further action research. Through purposeful inquiry, the teacher's role becomes one of active participation in the generation of new knowledge needed to make sense of science teaching and learning.

## ACTION RESEARCH: CRITICAL CONSTRUCTIVIST PERSPECTIVES

Teacher learning, when viewed from a constructivist perspective, requires opportunities for the teacher-learner to directly experience and critically reflect on the problems associated with science teaching and learning. Consider the following example:

Laboratory science typically involves procedures to be followed. But what happens when a chemical, specimen, tool, or sample is faulty? Most teachers quickly develop a substitute for that item. This type of personal research changes the classroom practice. Similarly, when a laboratory supply is low or not available, teachers often improvise. This improvisational action emancipates teachers from the traditional cookbook laboratory curriculum. Science teachers easily see this type of alteration to the curriculum as practical survival skills necessary for classroom teaching. But this example of action research falls short in its failure to critically examine the broader social and political forces that encompass the cultural context of science teaching and learning—a type of *critical constructivism* (Kincheloe, Steinberb, & Tippins, 1992). Critical teachers in the classroom recognize that science is not neutral, apolitical, or nonracist; they endeavor to confront dilemmas of science teaching and learning that are not always at the level of conscious attention.

Meaningful reflection and change can not occur unless the interests of those who are involved are heard, used, and acted upon. In this sense, action research must become critical. The core of critical action research supports the idea that learning occurs when teachers from self-critical communities of inquirers examine the culture of science teaching, teacher beliefs, and practices. Through critical action research, teachers seek to build on their own histories and intellectual interests in order to better understand the students' histories, concerns, strengths, and weaknesses. Cochran-Smith and Lytle (1993) state "when teachers do research, they draw on interpretive frameworks built from their own histories and intellectual interests." These interpretive frameworks can include reflective tools such as metaphors, metonymies, proverbs, and critical autobiographies. Far too often, the image of teachers engaged in classroom research is one that does not fit with the historically painted picture of teachers. The new image must consider the ways in which competing forces (e.g. power, race, class, gender, socioeconomic status, justice, and ethics) shape curriculum and instruction.

When teachers and their students engage in critical action research, two important questions emerge: (a) Whose interests are served by implementing action research? and (b) Who is in the best position to view, interpret, and implement the action research? On the one hand, everybody (students, teachers, schools, communities) profits from the shared knowledge generated by action research. At another level, when teachers and students collaborate in research, the results are shared but remain locally practical. Clandinin (1986) suggests that it is this locally produced knowledge that allows teachers and students to make decisions that directly affect their classrooms. Since action research does not originate from an anthropologist or university researcher, it's the native inhabitants—teachers and their students—who are in the better position to interpret their own experiences. As teachers reflect critically on their experiences, they learn to articulate ideas that emancipate them from the humanly-constructed social, political, and cultural distortions that frustrate and constrain self-understanding. Based on their research-driven analysis, teachers can then develop curricula for their school by researching problems and sharing their finds with other teachers.

## SUPPORTING TEACHERS AS LEARNERS

The creation of new knowledge about science teaching and learning is contingent on teachers having opportunities to engage in action research. In many schools, time for further learning, reflection, and analysis is a remote possibility amidst a pervasive crisis management atmosphere. Although action research is not meant to be an elu-

sive or elaborate process that may befuddle many, it does recognize that research is hard work. Anyone who has engaged in action research recognizes that "those conducting the research 'live' it as they think about whether their ideals match with their actions" (Manning & McLaughlin, 1995, p.7).

If schools are to find authentic and lasting ways to support teachers as learners, they must give science teachers time to talk about science teaching. This action research initiation phase, which provides opportunities for teachers to talk about science teaching, is perhaps the most important step in creating an environment in which teachers' research is held in high esteem. Science leaders should facilitate collaborative inquiry among teachers as they seek to identify problems and issues for study. In particular, they should encourage teachers to seek answers for questions that traditional empirical forms of research have often ignored. Reflective interviewing, analytical discourse, and graphic representation (e.g., webbing, concept mapping) are a few of the many strategies that can be used to assist teachers in identifying the issues to be studied. Teachers and science leaders should focus on personal knowledge, interests, and ideas as a basis for clarifying issues. As ideas develop, questions should be formulated to guide subsequent data framing and collection.

Emily F. Calhoun (1991) provides a framework for thinking about the phases involved in the implementation of action research:

(a) What to study? The topic needs to be personal. Researchers need to gather ideas from teachers' images, metaphors, metonymies, and interests.

(b) What data should be collected? How? When? How often? These questions should be addressed at the start or during action research. Data sources include tests, journals, logs, videos, interviews, archival evidence, surveys, and attestations.

(c) Framing and organization of data follows data collection.

(d) Data analysis.

(e) The final phase is the implementation of change in the classroom, schoolwide, and districtwide. The culmination of action research should be a written account in a peer journal.

These five phases represent a template from which teachers can construct their own action research.

Once the inquiry focus has been determined, collect and analyze existing data to determine if a change or study is necessary. Existing data may include conventional sources such as laboratory reports, test scores, grades, class enrollments, or lesson plans. These conventional sources of data can serve as a starting point for the development of more elaborate and creative data sources needed to uncover the hidden world of science teaching and learning. Action research, when grounded in a critical constructivist perspective, is not limited to educational questions that can be quantified. Kincheloe, Steinberg, and Tippins (1992) state:

> If we have any hope understanding the impact of schools on our students, the role schools play in the constructions of their consciousness, the relationship between schools and society, we will have to expand our research vocabulary to include both qualitative and quantitative methods (p. 238).

As data is framed and collected, teachers should analyze it for themes, assertions, ideas, patterns, and conclusions. The final phase of action research involves the implementation

of desirable changes, together with the corresponding creation of dialogue needed to unlock and change beliefs and practices.

## PROSPECTS AND IMPLICATIONS FOR SCIENCE TEACHER EDUCATION

The final consequence of action research is change. This change can appear in many forms, including professional growth, knowledge generation, and empowerment. The implications of action research must be explored, particularly in terms of its "continuing redefinition of teachers' knowledge base, the consideration of teachers' positions in the development of that knowledge base, and the development of different ways to explore practice" (Hamilton, 1995, p. 79).

When action research is initiated, the practice of teachers will change forever. This change may be reflected in teaching styles, instructional strategies, curriculum development, or organization of the learning environment. Teachers as critical action researchers bring their interpretations, histories, and personal understandings to the research context; in the process they empower themselves with the knowledge from their own perspectives that they will need to take ownership of the research process. When teachers have ownership in the sense-making process, they construct new courses of action based on their own ideas and those of their students.

If action research is to be considered as a valuable tool for initiating and sustaining change in science teaching and learning, then it must be included in preservice and inservice teacher education programs. Knowledge generated through action research stems from the experiences of teachers; accordingly, action research should be at the heart of collaborative school-university partnerships and science teacher education programs. In a context of co-reform, teachers as researchers can create a collaborative culture in which those learning to teach, those already teaching, and those working within teacher education reflect upon the complex dilemmas embedded within classroom practice. Science teaching and learning is ultimately transformed as prospective and practicing teachers engage in action research, question and reflect on their practices, test their assertions, and begin to generate their own emerging theories.

## References

Calhoun, E.F. (April,1991). *Action research for school improvement: The short no-frills course.* Paper presented at the annual meeting of the American Educational Research Association, Chicago.

Chein, I., Cook, S.W., & Harding, J. (1948). The field of action research. *American Psychologist, 3*(1), 43–50.

Clandinin, D. (1986). *Classroom practice: Teacher images in action.* London: Falmer Press.

Cochran-Smith, M., & Lytle, S. L. (1993). *Inside/outside: Teacher research and knowledge.* New York: Teachers College Press.

Corey, S. (1953). *Action research to improve school practices.* New York: Teachers College Press.

Dana, N.F. (1995). Action research, school change, and the silencing of teacher voice. *Action in Teacher Education, 16*(4), 59–70.

Dewey, J. (1993). *How we think: A restatement of the relation of reflective thinking to the educative process.* Chicago: D.C. Heath.

Grundy, D. (1987). *Curriculum: Product or praxis.* London: Falmer Press.

Hamilton, M.L. (1995). Relevant readings in action research. *Action in Teacher Education, 16*(4), 79–81.

Kincheloe, J.L. (1991). *Teachers as researchers: Qualitative inquiry as path to empowerment.* New York: Falmer Press.

Kincheloe, J., Steinberg, S., & Tippins, D. (1992). *The stigma of genius: Einstein and beyond modern education.* Durango, CO: Hollowbrook Press.

Lieberman, A. (1986). Collaborative research: Working with, not working on. *Educational Leadership, 43*(5), 28–32.

Lytle, S.L., & Cochran-Smith, M. (1992). Teacher research as a way of knowing. *Harvard Education Review, 62*(4), 447–474.

Manning, B.H., & McLaughlin, H.J. (1995). Editors comments. *Action in Teacher Education, xvi*(4), vii.

McKernan, J. (1987). Action research and curriculum development. *Peabody Journal of Education, 64*(2), 6–19.

National Science Teachers Association (1992). *The Content Core: A Guide for Curriculum Designers, 1.* Arlington, VA: Author.

Oja, S., & Smullyan, L. (1989). *Collaborative action research: A developmental approach.* London: Falmer Press.

Richardson, V. (1994). Conducting research on practice. *Educational Researcher, 23*(5), 5–10.

Schon, D.A. (1987). *Educating the reflective practitioner: Toward a new design for teaching and learning in the professions.* San Francisco: Jossey-Bass.

Sirotnik, K. (1988). The meaning and conduct of inquiry in school-university partnerships. *School-university partnerships in action : Concepts, cases, and concerns,* by K. Sirotnik & J. Goodlad (Eds.), (pp. 169–190). New York: Teachers College Press.

Tobin, K., Tippins, D., & Hook, K. (1994). Referents for changing a science curriculum: A case study of one teacher's change in beliefs. *Science and Education, 3*(3), 246–265.

Vitale, M.R. & Romance, N.(1994). Evolution of a collaborative school research project as a prototype environment for empowering teacher action research: Results and implications. *Florida Journal of Educational Research, 34*(1),43–50.

## Author Note

**Deborah J. Tippins** is a professor of elementary education and science education at the University of Georgia. She has written numerous articles and grants in science education, and has won numerous awards for her presentations and publications, including NARST's Early Career Research Award. She is currently on the board of NSTA.

**William R. Veal** is a graduate student in science education at the University of Georgia. He has taught physical science, chemistry, physics, and advanced placement chemistry at the secondary level for four years. He has published research articles in chemistry and education. His interests include action research, teacher education, and pedagogical content knowledge.

# Research in the Classroom

Michael P. Marlow
Stacey E. Marlow

As Jennifer drew a sample of water into a plastic syringe, Kate looked on awaiting her chance to balance the water's acid level by squeezing seven drops of baking soda into the 29-gallon tank. Not many third-graders have the opportunity to study an endangered species as these students can. The students are the youngest class involved in the Lake Victoria Endangered Lake Fish Project (ELF). They are caring for six Hippo Point Salmon Cichlids from East Africa. Twenty-seven other classes, grade 4 through 12 in ten different school districts, are also participating in this research project and have the Hippo Point Salmon Cichlids in their classrooms. The purpose of the project is to incorporate a real-life research component into district science programs. Research opportunities vary from individual, to group, to classroom-size projects. Most activities are predetermined as part of an articulated research agenda but school-initiated research is encouraged, as long as it follows the research protocol established for the program. School participation is contingent upon their agreement to abide by all procedures and protocol for handling the fish stocks and experimentation. Procedures and protocol were developed by a regional committee that oversees the project. The ELF program is a collaboration among the Lake Victoria (Africa) Society, a university, a group of area school districts, and a county Math/Science Center to raise and investigate Lake Victoria endangered fish. Internet connection allows each classroom to communicate research findings to other classrooms as well as to university and aquarium researchers. Although there is a specific research agenda that all classes follow, the project crosses disciplines. As Jennifer's teacher, Mrs. Hall explained:

> I use the project to teach a variety of subjects from math to history. Students need to know how all different subjects are interrelated. Making the school experience real—that's what we're all about. Students apply math skills by measuring food and water amounts. History and social skills come into play by learning about the African continent and how the endangered fish have affected the people of the region. English skills are used when students record what they have learned about the fish (Riley, 1994).

Teachers across the country are beginning to explore alternative methods of instruction, such as the Lake Victoria Project, that incorporate student research into their science teaching. Rather than relying solely on textbooks or lecture-based instruction, teachers are attempting to provide more opportunities for students to engage in authentic tasks, for example, analyzing the water quality of a local river and/or doing research on local impacts of a specific insect pest. This long-term, collaborative, concept rich, process- and product-intensive activities produce high student motivation and involvement while improving learning of both process and content. While many teachers recognize the need to move in this direction, how to plan, set up, and manage such a project is beyond their present professional preparation.

## WHY INCORPORATE RESEARCH IN THE CURRICULUM?

While there is no consensus on a single definition of scientific literacy, many agree that it encompasses three aspects of science. They are product, process, and values. Product is the content information of the sciences. It also includes knowledge of the nature of science, particularly its methods of investigation. The processes of science are the thinking skills used to solve problems and to conduct investigations, while scientific values are the beliefs and attitudes, such as logic and curiosity, that underpin the practice of science.

All children express a need to interact with adults and with each other, with materials, and with their environment in ways that help them to make sense of their experiences and the world around them. Generally, however, teachers spend too much time on paper-and-pencil tasks and not enough time on hands-on learning that engages the total individual. It is essential for teachers to understand how children learn and to offer them rich, creative experiences and opportunities to apply what they have learned in other contexts as well as in the classroom. Clearly, no single approach can

meet the needs of all students. We must begin to integrate the most effective ways to teach and learn into our schools.

Constructivism is a theory of knowledge used to explain how we know what we know. Learners use all their senses to interact with the environment and construct meaning. According to this theory, knowledge resides in individuals and cannot be transferred intact from the mind of the teacher to the minds of students. Rather, students make sense of what is taught by trying to fit it within their experience (Lorsbach & Tobin, 1992). From the constructivist perspective, science is a process that assists students in making sense of the world. As Lorsbach and Tobin write:

> Using a constructivist perspective, teaching science becomes more like the science that scientists do—it is an active, social process of making sense of experiences, as opposed to what we now call 'school science.' Indeed, actively engaging students in science (hands-on, minds-on science) is the goal of most science education reform(p. 2 ).

The teacher's role within this approach is not the lecturer of the traditional science class, but rather a facilitator of students' learning and an active participant in using and discussing science.

Lederman (1992) found that :

> In general, the classes with the most effective teachers were typified by frequent inquiry-oriented questioning, active participation by students in problem solving activities, frequent teacher-student interactions, infrequent use of independent seat work, and little emphasis on rote memory/recall. With respect to classroom climate, classes of the more effective

teachers were more supportive, pleasant, and risk free with students expected to think analytically about the subject matter presented (p. 348).

Problem identification, analyses, and problem solving abilities leading to scientific literacy are not advanced through rote learning; in fact, its very nature requires a more open approach to critical thinking. Rutherford and Ahlgren in *Science For All Americans* (1990) state:

Students need to have many and varied opportunities for collecting, sorting and cataloging, observing, note taking and sketching; interviewing, polling, and surveying; using hand lenses, microscopes, thermometers, cameras, and other common instruments. They should dissect; measure, count, graph, and compute; explore the chemical properties of common substances; plant and cultivate; and systematically observe the social behaviors of humans and other animals (p. 188 ).

The type of classroom advocated here is one that supports these ideas through collaborative work, discussion and sharing of ideas, mutual respect for each learner's approach, and students' sense of ownership of their work.

## EDUCATIONAL APPROACH

The incorporation and implementation of student research into the science curriculum require strategies for developing research skills. One component of any research inclusion program should be to identify and define the basic concept of experimental design, which includes hypotheses, dependent and independent variables, control groups, constants, and repeat trials. This component of the model is constant regardless of the research topic.

The usual method for introducing research to students is to begin by teaching the steps of the scientific method. Typically, five or six major steps are listed and defined. Students are then instructed to identify a project topic, select a specific problem, and design an experiment. This approach has many problems. First, the scientific method as taught in many classrooms describes the way experiments in research are reported, not how they are generated. Second, the instructional sequence described in the method usually does not sufficiently develop the basic concepts of experimental design: hypotheses, dependent and independent variables, control groups, constants, and repeat trials. Third, the students often lack sufficient background knowledge and experiences that allow them to self-select a research topic and carry it to a meaningful conclusion.

An effective method for teaching students to design their own research is to begin with a predetermined activity that allows students to manipulate materials and see the results. In the Lake Victoria project each group begins by learning about the nitrogen-ammonia cycle and the potential effect of ammonia imbalance on fish and then proceeds to balance the host aquarium. After newborn fish are placed in the aquariums, all classes participate in the same research agenda that collects data on the growth of the cichlids over the first three months of the project. Groups of aquariums have different feeding schedules, amount of food, and water change schedules, thus introducing variables. Data is collected, placed on graphs and shared by electronic communication with other classes in the project. Because each classroom has the same size and model aquarium and filter systems and were set up the same way, these variables are removed. Through these activities, students begin understanding research protocol and techniques. As the project continues, classes begin to discuss how they might design their own research project using the fish. For example, most students had observed early on that the fish changed their color when interacting with other fish in the tank. By observing these interactions, students developed experimental designs that were clarified through effective teacher questioning and class discussion. Formal concepts and terms that emerge from these discussions held meaning for students because they were based upon real observation. Research activities evolved that systematically charted color change and location. One fifth-grade class charted all color interactions when one, two, or more fish left their hiding tubes, compiling six months of detailed data. The student's motivation was exceptionally high.

## APPLYING RESEARCH CONCEPTS

Knowledge of the basic concepts of research design and the ability to apply these concepts represent two different levels of learning. Students may be able to describe and/or define the steps or identify examples in a classroom investigation yet fail to apply the concepts when designing their own research. Students' ability to apply the basic concept of research design is enhanced by thoughtful written descriptions of the research before and after conducting it. Through writing, students practice identifying and diagramming the basic aspects of the research. Discussions of variables provide a basis for discussing typical design flaws and lead to methods of improvement. In the Lake Victoria project students keep logs, have teacher- and student-led classroom and small group size discussions, and have ongoing contact with other classes over e-mail. Classroom generated research is written up before implementation and presented to an outside advisory group for comment.

## GENERATING RESEARCH IDEAS

In a typical classroom, students are often asked to generate experiments through discussion within cooperative groups of peers or select from a list provided by the teacher. Often these topics are unconnected with anything that has previously taken place in the class. Students placed in this situation will frequently have a difficult time generating an original experiment. It doesn't matter how well the student has memorized the scientific method, or even the definitions of variables, controls, and repeat trials. The result is generally the same—disinterest, frustration, and disconnected thoughts. Science teachers need to have their students first participate in a planned research activity with all aspects of the project clearly identified and discussed, then students can begin developing possible variations of the research theme. They can begin by asking four questions: (a) While doing the initial research did I notice any activity or event that generated new questions and needs further observation or experimentation? (b) How can I change the set of variables to observe or affect this event? (c) What materials would be needed for conducting this new experiment? (d) How can I measure or describe the response of the subject to the change?

## TABLES AND GRAPHS

Researchers communicate information in specialized ways including data tables and graphs, as well as through speaking and writing. Unfortunately, many science classes do not emphasize data presentation and graphing skills. Initially students should enter numbers into prepared data tables and plot points on pre-constructed grids designed by the teacher. This gives students concrete examples of design. Taken alone, however, such activities teach students only how to record data, not how to communicate their meaning. Students learn communication skills by self-designing data tables, constructing graphs, and formulating sentences that communicate their findings. In the Lake Victoria project, students use pre-constructed graphs and data tables for the first round of activities but develop their own forms for the classroom-initiated research. In the elementary classrooms students enter the data as a group on large posterboard sized charts and graphs that are designed by the entire class through group work followed by whole class discussion. These steps are essential for development of higher-level thinking skills.

For students to be scientifically literate, they must be able to analyze and interpret data. Strategies must be used to teach older students to distinguish among major categories of data and to identify levels of measurement. High school research requires that additional time and effort be spent on issues such as communicating descriptive statistics, determining statistical significance, designing complex experiments, and presenting student research.

## REPORTING FINDINGS

Getting students to write a detailed procedure for a scientific investigation is a major challenge. Involving all students in the data collection, maintaining descriptive logs to which all students in the class contribute, and allowing students to prepare a class report of the research help model this component. Classroom discussions about what should go into the report and how it should be expressed will allow students to contribute information and ideas that combined will result in a more detailed report. Observing and actively participating in this experience will help students recognize what their individual work should include. Letting the students interact and review one another's writing will further enhance the final individual product. After students have written individual draft reports, they work in small groups to analyze one another's efforts for missing information.

A research report generally answers six questions: (a) What was the purpose of the research? (b) What were the major findings? (c) What research hypothesis is supported by the data? (d) How did your findings compare with other researchers' in the project's findings? (e) What possible explanations can you offer for the similar/dissimilar findings? (f) What suggestions do you have for further study and for improving the experiment?

## INTEGRATION ACROSS THE CURRICULUM

Most student understandings can be enhanced by connecting concepts across a number of disciplines. Effective learning, however, often requires more than simply making multiple connections of new information to existing ideas or beliefs. Learners need to be provided opportunities that allow them to confront existing naive conceptions or misconceptions through a variety of learning experiences that result in the restructuring of their thinking in ways that help them better understand the world (American Association for the Advancement of Science, 1989).

Enrichment experiences in the Lake Victoria project are numerous but generated individually by class or teacher. Therefore some classes expand their understandings beyond those of other classes. This was especially true in the elementary classes. Teachers, in an attempt to help students better understand the impacts of an endangered species on its local environment, developed social studies units that centered on the countries that surrounded the lake. In direct student-to-student e-mail contact between project classrooms and classrooms in countries surrounding the lake, they learned that the cichlids had been a food source for people living around the lake and that the protein supplied by this food source had not been replaced in the people's diet. They learned about how the lake became polluted and how the introductions of a new fish species, the Nile perch, had decimated the native cichlids. They also exchanged information on general lifestyles that generated units on music and art from the African region, helping students recognize that real people were part of this problem. The use of technology provided many of these opportunities to learn. Also, a great deal of math was directly imbedded in the project, and many teachers used the project to help students better understand additional math concepts.

The high school teachers were not as open to this type of enrichment, feeling that time was not available to move much beyond the established science protocol. They generally were open to establishing e-mail connections with additional researchers and science resource specialists but were less interested in social science issues or in collaboration with other curriculum area teachers within their school. The richer curriculum integration developed at the elementary level, with its resulting greater student involvement, would seem to indicate that future implementation of this type of project should include designing strategies to encourage and assist more curriculum integration on the secondary level.

## TEACHER PROFESSIONAL DEVELOPMENT

Placement of equipment and suggestion of a research topic alone will not generally result in a successful project. This is especially true in elementary classrooms. A professional development plan must accompany the new project. This plan must contain ongoing support for the teacher both in the form of a series of planned meetings and workshops, as well as continuous access to a resource/information system related to the project. The system developed for the Lake Victoria project involves a series of meetings prior to the placement of the fish in the classrooms. Sessions are presented on general aquarium management, water chemistry in an aquarium, history of the research topic, introduction of the researchers that would act as support, research techniques in general, and the specific research protocol for this project. E-mail connections are established

for teachers to communicate with the researchers and other teachers in the project. Bimonthly meetings are held to discuss problems, results, and new information.

## CONCLUSION

The type and range of research activities, and the associated workshops and academic preparation should be chosen for their impact on school and district goals relating to successful student learning. Based on the implementation of the Lake Victoria experience, we have identified a number of areas that we feel should be incorporated into any student research project. These areas are:

• All research emphasizes active, hands-on involvement of students.

• Sessions emphasize the inclusion of all students rather than selecting out or tracking.

• Extended teacher professional development components are part of all programs.

• Programs should, where appropriate, contain "integration across the curriculum" components.

These areas are based on the educational assumptions that:

• All children possess several types of intelligence.

• Students learn in individual ways and should have a variety of learning.

• Students should be actively engaged in learning that is meaningful to them.

• Student growth and program effectiveness should be assessed in a variety of ways.

• The learning environment should incorporate, and be based upon, established research and sound learning theory.

• Technology should be used as an enabling tool that supports the learning process.

With these assumptions in mind, students should be able to:

• Demonstrate the ability to approach a problem/project logically.

• Exhibit a positive attitude toward learning.

• Demonstrate the ability to work with others in completing a task.

• Communicate effectively.

• Display the ability to seek out and evaluate information.

• Practice self-direction and good study/work habits.

• Function with integrity.

• Exhibit an understanding of self and others.

Authentic research programs such as the Lake Victoria project represent a radical departure from traditional science curriculum and instruction for many teachers. In order to successfully implement student research in the classroom, teachers must be knowledgeable about not only the subject matter but the research process and the relevant learning theories. Furthermore, they must have a clear understanding of the educational assumptions underlying the use of research in the classroom. But teachers do not stand alone in the introduction of student research in the classroom. Schools and school districts contemplating the incorporation of authentic research programs into their curricula can assist teachers by providing adequate infrastructural support in the form of resources and materials. They must

be prepared to make a commitment to providing the necessary resources and continuous staff development that will support the efforts of teachers in the classroom.

## References

American Association for the Advancement of Science, Project 2061. (1989). *Science for All Americans: Project 2061*. Washington DC: Author.

Lederman, N. (1992). Students' and teachers' conceptions of the nature of science: A review of the research. *Journal of Research in Science Teaching, 29*, 331–359.

Lorsbach, A., & Tobin, K. (1992). Constructivism as a referent for science teaching. *NARST: Research Matters—to the Science Teacher,* No. 30.

Riley, K. (1994, February 21). Third graders help save a species. *The Jackson Citizen Patriot,* p.1

Rutherford F.J. & Ahlgren, A. (1990). *Science for all Americans.* New York: Oxford University Press.

## Author Note

**Michael P. Marlow** is currently associate professor of science education at the University of Colorado at Denver. He developed the Lake Victoria Project in 1993–94 while Director of The Jackson County Mathematics and Science Center in Jackson, Michigan. This is one of a number of student research projects he initiated through the Center. Prior to that, he was a middle school science teacher.

**Stacey E. Marlow** is an assistant professor in educational leadership at the University of Hawaii at Manoa. She has participated in numerous science curriculum development and implementation projects and has written articles and presented at national conferences on the topic. She began her career as an elementary teacher in Michigan and later was a district curriculum consultant. She received the 1995 University of Hawaii Regents Medal for Teaching Excellence.

The Lake Victoria Project has completed a second year in a number of school districts in the Jackson, Michigan, area. Information can be obtained from the Jackson County Mathematics and Science Center, 6700 Browns Lake Rd, Jackson, Michigan 49201 or Robert Stowe at Grass Lake High School, 1000 Grass Lake Road, Grass Lake, Michigan 49240. The Lake Victoria Society, whose director is Richard Preuss, can be contacted at Lake Victoria Society, Inc. 1651 Haslett Road, Hasslet, Michigan, 48840.

# Authentic Assessment: Vehicle for Reform

Glenda Carter
Sarah B. Berenson

Misconceptions research of the past two decades produced evidence that students lacked understanding of even the most fundamental science concepts. One response to this finding was to initiate a reform of curricula. Another response was to acknowledge the need for assessment measures that would elucidate levels of conceptual understanding. *Authentic* or *alternative* assessment became popular terms to describe measures that could accurately reflect students' conceptual understanding of science and measure their ability to solve problems as a result of their understanding. However, before the educational community had reached consensus about what constituted valid and reliable authentic assessment, these assessments were mandated at statewide as well as local levels. While there were certain advantages to this approach, there were many drawbacks. The greatest advantage was in the immediate focusing of both monetary and human resources toward the goal of improving assessment. The drawbacks included implementation of some assessments that were no better than the ones they were replacing and frustration of students, parents, and teachers who were ill prepared for such a change. It was particularly stressful for teachers who were asked to prepare students for a type of assessment that they themselves knew very little about, while being held ultimately accountable for how well their students did on this new type of assessment.

Therefore one of the first demands on the science education community was to provide professional development programs for teachers in authentic assessment. Figure 1 shows students' responses to an authentic item. It is apparent that these kinds of assessments demand a new response repertoire from science teachers. Authentic assessments require a level of analysis and interpretation far beyond traditional testing. Indeed, teachers become classroom researchers within the active roles of teaching and learning. A professional development program for teachers can assist them in (a) developing authentic assessment tasks, (b) reflecting on the meaning and implications of students' responses, and (c) designing instructional interventions to improve students' understanding. In other words, teachers become action researchers within their own classrooms.

## MODEL OVERVIEW

A constructivist approach for this model was an obvious choice for two reasons. First, the underpinning theory of the reform

It is hot in the summer because as the earth rotates and revolves different parts of the earth face the sun, and in the summer our part of the earth is facing the sun so it is hotter then. In the winter our part of the earth is on the opposite side of the sun so it becomes colder.

It's cold in the Winter because half of the earth is tilted away from the sun and not as much heat is on the side where the earth is tilted away from it. The other half is tilted to the sun and it's getting most of the heat.

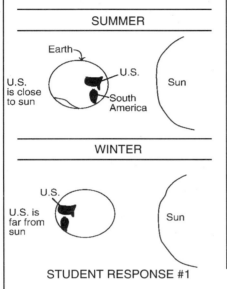

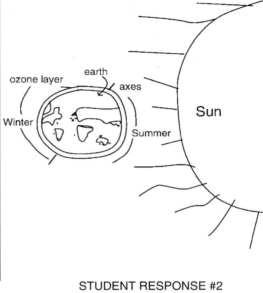

STUDENT RESPONSE #1

STUDENT RESPONSE #2

movement is based on the tenets of constructivism. Second, research on successful models of professional development indicates that teachers construct their knowledge about pedagogy and related content primarily through interactions with their students and reflective interactions with their colleagues (Wood, Cobb, & Yackel, 1991). A straightforward mechanism for incorporating these reflective processes is action research.

The three phases of action research are:

1. *Reconnaissance*—the identification of problem and recognition that change must occur.

2. *Plan*—the development of a plan of action including changes in practice, intended effects, and data collection.

3. *Reflection*—the reflection on what has or has not happened (Perry-Sheldon & Allain, 1987).

This model of action research is cyclical in nature with the reflection stage giving rise to the next reconnaissance phase. This cyclical model has two advantages. First, it provides the opportunity for teachers to master and refine the knowledge and skills necessary to implement authentic assessment through repeated efforts and cycles of reflection. Second, teachers are more likely to change classroom practices when the change is catalyzed by results based on teachers' research in their respective classrooms (Simmons, 1984).

In the first phase of the professional development program, teachers are introduced to various formats of authentic assessment. They examine sample student work and review concepts they will be teaching before writing authentic tasks related to those concepts. With their workshop colleagues, teachers review and develop an implementation plan for their assessments. After six to nine weeks in the classroom, teachers

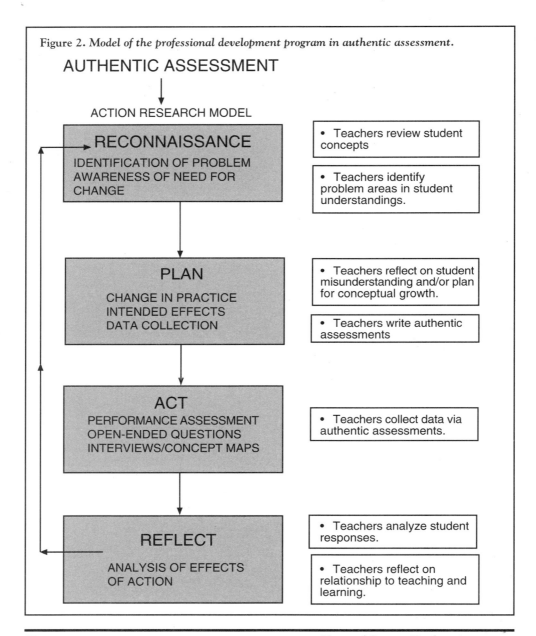

Figure 2. *Model of the professional development program in authentic assessment.*

AUTHENTIC ASSESSMENT

ACTION RESEARCH MODEL

**RECONNAISSANCE**
IDENTIFICATION OF PROBLEM
AWARENESS OF NEED FOR
CHANGE

- Teachers review student concepts
- Teachers identify problem areas in student understandings.

**PLAN**
CHANGE IN PRACTICE
INTENDED EFFECTS
DATA COLLECTION

- Teachers reflect on student misunderstanding and/or plan for conceptual growth.
- Teachers write authentic assessments

**ACT**
PERFORMANCE ASSESSMENT
OPEN-ENDED QUESTIONS
INTERVIEWS/CONCEPT MAPS

- Teachers collect data via authentic assessments.

**REFLECT**
ANALYSIS OF EFFECTS
OF ACTION

- Teachers analyze student responses.
- Teachers reflect on relationship to teaching and learning.

reconvene bringing the results of assessment tasks and their reflections about the results of these assessments. At this time teachers enter the first stage, the *reconnaissance stage*, of the action research model (Figure 2). By sharing their student data with colleagues, teachers identify problem areas in conceptual understanding. Then alternate intervention strategies are developed along with the plan for the implementation and assessment of these strategies. Returning to the classroom, teachers are prepared to intervene with their students and collect data by means of authentic assessment tasks. Teachers continue to come together throughout the school year to share their reflections. As they analyze results of these assessments and plan instructional strategies, they recirculate through the action research model.

As the model indicates, teachers use authentic assessments to gain a clearer view of the students' understandings, and they are reflecting on teaching practices that will affect their students' understanding. Action research and authentic assessment bring about instructional changes in the classroom. But the changes are initiated by the teachers themselves in response to the needs of their students. By using authentic assessments not only are teachers convinced of the need for reform, but they become convinced of their ability to affect such change in their classrooms.

## FORMATS FOR AUTHENTIC ASSESSMENT

There are several forms of authentic assessment currently being used to measure understanding of science concepts, including open-ended questions, performance assessment, concept maps, interviews, and portfolios. An intended outcome of this particular model of a professional development program is to bring all participants to the awareness level (recognition) for these formats of authentic assessment. We've found that the personal preferences of the teacher partici-

pants and their particular student populations will determine those formats in which the teachers will elect to move to either the implementation level or expert level.

### The Open-Ended Question

One of the richest formats of authentic assessment is the open-ended question. By definition open-ended questions are those that have more than one correct answer or, when process is being evaluated, have more than one acceptable way to arrive at the correct answer. Open-ended responses assist teachers in more clearly understanding how students think, the prior knowledge students bring to a subject, and what understandings are gained during the instructional process. Teachers who frequently use open-ended questions indicate that students become more proficient in thinking divergently and creatively. Additionally, as students are encouraged to share and negotiate answers, teachers become participants in the educational dialogue rather than the authorities who dispense right answers (deLange, 1993).

One critical task of the professional development program is to assist teachers with writing good open-ended questions. Although there are some published sources of open-ended questions (Carter & Henderson, 1994), there is an expressed need for teachers to be able to write their own items based on their conceptual targets and the level of their students' responses. A guideline (Figure 3) for assisting teachers in writing these tasks was developed, piloted, and reported by Berenson and Carter (1994).

In addition to this guideline, participants of the professional development programs developed criteria to guide the quality of the open-ended questions they are producing and implementing. After implementation, the responses to questions are analyzed to determine if each question:

• provides insight into student learning.

**Figure 3.** *Guideline for writing open-ended questions.*

\* WRITE \* EXPLAIN \* DESCRIBE \* TEACH \* CONSTRUCT \*

### STORY

1. Students write a story problem or a story for science that describe:
   - two related variables: time and distance, time and temperature, distance and weight
   - cycles in nature
   - similarities and differences among natural objects

### OPINIONS

2. Students explain their opinions about a ... statement:
   - "Some students think that [misconception or ambiguous statement]. What do you think? Explain your thinking in words and pictures"

   i.e. Some students think that a full moon occurs when the moon passes between the earth and the sun. What do you think?

### DESCRIBE

3. Students describe in words/pictures a scientific phenomena.
   i.e. Describe in words and pictures how a tornado is formed.

### TEACH

4.1 Teach another student how to ...
   i.e. Explain how you could use our schoolyard to teach another student about erosion.
4.2. Describe the error that this student made on the ... test.
   i.e. Don stated that mixing sugar and water was a chemical change. How would you help Don understand his error?
4.3. How would you use this (equipment) to teach this [concept] to younger children.
   i.e. Describe how you would use a graduated cylinder to teach another student the concept of volume.

### CONSTRUCT

5.1. A graph
   i.e. Construct a graph that illustrates the results of your experiment.
5.2. Demonstration
   i.e. How would you demonstrate evaporation to a younger student?
5.3. Model
   i.e. Design a model for protecting the integrity of our salt marshes.
5.4. A table
   i.e. Construct a table to show the data that you collected on rainfall
5.5 A diagram
   i.e. Draw a diagram that illustrates the water cycle.
5.6. An experiment
   i.e. Design an experiment to determine which soil is best for growing tomatoes.

- promotes higher levels of thinking.
- stimulates interest.
- elucidates different levels of understanding.
- measures conceptual understanding.
- allows for a variety of acceptable responses.
- is clearly worded.
- does not lead students to a specific answer.

## Performance Assessment Tasks

A performance assessment task is one that includes students' manipulation of some material to produce a product that illustrates conceptual understanding. This product can be written, pictorial, kinesthetic, or a combination of modes. The performance assessment task either is open-ended or has a discrete answer. Performance assessment tasks are usually selected to assess achievement in

Figure 4. *Guideline for writing performance assessment tasks.*

### PERFORMANCE-BASED ASSESSMENT IN SCIENCE
#### Specification sheet

**Curriculum Alignment:**

1.  Topic ("Big Idea")

2.  Goals/Objectives

     A. Content

     B. Process

3.  Purpose of activity/expected outcome:

4.  Skills necessary to complete assessment:

==================================================================

**ITEM DESCRIPTION:**

1. Problem statement (Student task)

2. Item Logistics

     A. Materials needed:

     B. Projected Student Completion time:

     C. Space requirement:

     D. Advanced Preparation required:

==================================================================

**STUDENT ANSWER SHEET FORMAT:**

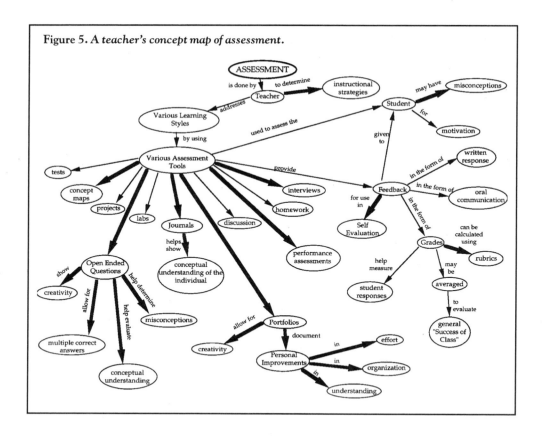

**Figure 5.** *A teacher's concept map of assessment.*

using the process skills. *Science Process Skills* (Ostlund, 1992) provides an excellent resource and model for designing such tasks and includes over 60 such tasks on a variety of levels. Teachers generally do not have a difficult time implementing these tasks if a source of ideas is provided. The greatest impediment to implementation is in the management of the performance centers and time needed to assemble the equipment. A format to assist teachers with producing their own performance assessment tasks is located in Figure 4 (Scherr-Freedman, 1994).

## Concept Maps

A concept map provides a visual portrayal of an individual's mental representation of a concept (Novak & Gowan, 1984). Therefore concept maps are useful in providing both teachers and students with an under-

standing of prior knowledge and the conceptual gains that are made during a unit of study. Given a specified target concept, individuals construct their concept maps by writing down related concepts and drawing lines to represent linkages between and among concepts. Each linkage is labeled to indicate the individual's understanding of the relationship between the concepts.

During an early part of the professional development program, teachers are introduced to concept maps. After they are comfortable with the technique, they construct their own concept maps of assessment. At the end of the workshop participants either redraw their concepts map or make any desired changes in a different color ink on their original concept map. Figure 5 illustrates a teacher's pre- and post-concept map on assessment. Heavy lines indicate post-workshop changes. It is obvious that the assessment repertoire of the teacher improved as a result of the workshop. The most significant change is the identification of assessment tools as important feedback mechanisms for the determination of instructional strategies.

Articles by Vargas and Alvarez (1992) and Roth (1992) provide teachers with additional information on how to begin using concept maps as well as suggestions for incorporating concept maps in science instruction and assessment.

*Interviews*

Many teachers informally use interviewing on a daily basis without being aware of this process. They may question a disconcerted student to uncover the source of puzzlement, or they may interact with cooperative groups of students to probe understanding. When they are assessing student understanding for the purpose of instructional intervention, they are implementing the interview as a type of formative authentic assessment. Formal interviews can also be done, and those teachers who are willing to sacrifice the time at the beginning of a unit to probe a few students' ideas about an upcoming topic often find that it is quite revealing and often elucidates misunderstandings that can interfere with students' learning. Several articles have done an excellent job of reviewing key points and making suggesting for incorporating the interview into classroom instruction (Long & Ben-Hur, 1991; Peck, Jencks, & Connell, 1989; Schoen, 1979). A guideline (Berenson & Carter, 1995) for developing an interview is illustrated in Figure 6.

## Portfolio Assessment

The portfolio is a repository of selected works that illustrates growth in understanding and achievement over a specified period of time. Portfolio assessments may include any or all of the other types of assessments.

The work that is included in a portfolio is usually selected with input from both the student and the teacher. Although there is no one right way to do a portfolio, student reflections on their work are always an integral ingredient. They may be asked to discuss or exhibit work that represents their best effort, or they may be asked to include work that they feel needs improvement. Students may choose work from the beginning of the year and from the end of the year to illustrate progress they have made. Concept maps, journal writing, videotapes, projects, and responses to open-ended questions can be included in the portfolio to document changes in breadth and depth of student understanding.

Management of materials is a critical issue with portfolio assessment. Although there are electronic storage devices on the market, many teachers do not have access to this type of equipment. The intended purpose of the portfolio and the intended audience for the portfolio determine the best way to manage materials. Storing a working portfolio within the classroom in boxes or in binders can assist students with keeping up with papers until the time comes to prepare the portfolio.

One primary advantage of using classroom portfolios is to give students the opportunity to be a part of the evaluative process. Using portfolios as the focus for student-led conferencing at the end of each grading period or semester reinforces the students' responsibility for learning and for demonstrating their learning.

Having teachers keep a portfolio on assessment to document their growth and progress in understanding throughout the professional development program best demonstrates the process. Additionally, both *The Science Teacher* and *Science Scope* have published articles in the past few years that teachers find helpful as references for begin-

**Figure 6.** *Guideline for developing an interview for assessment.*

### Checklist for Interview Assessment

- Topic_____

  Targeted concepts:

- Results of written preassessment/classroom observations

  What characteristics of the concept make it difficult for students?

  What important subconcepts are necessary for understanding?

  What prior misunderstandings do students bring to the subject area ?

- <u>Interview setting</u>                    • <u>Recording method</u>

  ___Individual                              ___Checklist/written observations

  ___Small group                            ___Audiotape

  ___Whole group                            ___Videotape

- <u>Interviewees</u>

  ___Volunteers

  ___Determined by results of written preassessment

  ___Determined by classroom observations

  ___Students selected representing range of conceptual achievement

- <u>Opening question</u>

    "Can you explain …"

    "What do(did) you mean by …"

    "How would you …"

    "Tell me everything you know about …"

    "How would you explain … to another student"

- <u>Possible follow-up questions</u>

    "Tell me more about …"

    "Explain using an illustration/this picture/these instruments..."

    "Show me …"

    "What if …"

    "How are … related?"

ning this task (Collins, 1992; Hamm & Adams, 1991).

The Association for Supervision and Curriculum Development (Kiernan, 1991) has produced a series of videotapes on as-sessment including the portfolio format. Written materials with suggestions and guidelines for professional development programs on authentic assessment are included with the videotapes.

## IMPLEMENTING AUTHENTIC ASSESSMENT

Because teachers want to understand what their students are learning, they quickly become enthusiastic about the possibilities of authentic assessment. But successful implementation of authentic assessment must also include parents and students. The science education community along with the teachers who incorporate authentic assessment into their classrooms must be sensitive to parent and student needs and their responses to this new form of assessment. Students who have been successful with traditional forms of assessment may understandably be resistant to change. Parents whose children have been successful with traditional assessment often fear that authentic assessment tasks will lower their children's grades or academic standing. Most parents have had no experience with authentic forms of assessment and may fear its impact on their children's learning. Parents who are partners in this new approach from the beginning are seldom resistant to these changes. Just as with the teachers, parents and students who are given the opportunity to experience authentic assessment in a low stakes setting are more likely to become advocates for change. For instance one teacher, interested in moving toward portfolio assessment, used portfolios in addition to traditional assessment for the first grading period. Then she invited parents to an open-house where students showed their parents a sample of their portfolio work. The response of the parents was unanimously positive to this alternative documentation of their children's progress. After this overwhelmingly positive response, the teacher was able to use portfolios for the remainder of the year in place of more traditional forms of assessment.

## TEACHERS AS LEADERS IN THE REFORM MOVEMENT

Changing assessment is an integral part of the reform movement. Students take their cue about what they are expected to learn from the type of assessment that is administered. If students are primarily assessed by multiple choice and true-false questions, then factual recall remains the focus of the science curriculum. Using authentic assessments in the classroom can refocus teacher and student efforts toward conceptual understanding. Additionally, teachers involved in implementing authentic assessment become aware of their students' and their own conceptual understandings, and acquire a greater understanding of how teaching and learning are related.

One of the glaring errors of the 1960s reform movement was the belief that a foolproof curriculum, independent of the classroom teacher's knowledge and pedagogy, would bring about sweeping changes. Leaders in the science education community can offer teachers the opportunity to look at the conceptual development of their students through authentic assessment. This will serve as the keystone to a professional development program for catalyzing reform. If assessments are the basis for educational decisions, reform in educational practices can be made by targeting the results of assessment.

## References

Berenson, S., & Carter, G. (1994). Writing open-ended science problems. *Science Educator, 3,* 23–26.

Berenson, S., & Carter, G. (1995). *Alternative assessment: Practical applications for mathematics and science teachers.* (Available from the Center for Research in Mathematics & Science Education, Box 7801, North Carolina State University, Raleigh, NC 27695–7801.)

Berenson, S. & Carter, G. (1995). Changing assessment practices in science and mathematics. *School Science & Mathematics, 95* (4), 181–186.

Carter, G., & Henderson, N. (Eds.). (1994). *Open-ended questions: Alternative assessment for use with North Carolina Standard Course of Study* in Science, Grades 4-6. CRMSE. (Available from the Center for Research in Mathematics & Science Education, Box 7801, North Carolina State University, Raleigh, NC 27695–7801.)

Collins, A. (1992). Portfolios: Questions for design. Science Scope, 15, 25–27.

deLange, J. (1993). Assessment in problem-oriented curricula. Assessment in the mathematics classroom, by N. Webb & A. Coxford (Eds.) (pp. 197-208). Reston, VA: National Council of Teachers of Mathematics.

Hamm, M., & Adams, D. (1991). Portfolio assessment: It's not just for artists anymore. The Science Teacher, 58, 18–21.

A Handbook of Performance Activities: Science/ Social Science Grades 7–12. (1993). New York: Harcourt Brace Jovanovich.

Kiernan, L. (Producer). (1991). Redesigning assessment [videotape series]. Alexandria, VA: Association for Supervision and Curriculum Development.

Long, M., & Ben-Hur, M. (1991). Informing learning through the clinical interview. Arithmetic Teacher, 38, 44–46.

Novak, J. D., & Gowan, D. B. (1984). Learning how to learn. Cambridge, England: Cambridge University Press.

Ostlund, K. (1992). Science process skills. Menlo Park, CA: Addison-Wesley.

Peck, D., Jencks, S., & Connell, M. (1989). Improving instruction through brief interviews. Arithmetic Teacher, 37, 15–17.

Perry-Sheldon, B., & Allain, V. (1987). Using educational research in the classroom. Phi Delta Kappa Fastback, 260.

Roth, W-M. (1992). Dynamic evaluation. Science Scope, 15, 37–40.

Scherr-Freedman, D. (1994). Performance-based assessment in science: Specification sheet. (Instru-

ment available from Technical Outreach for Public Schools Project, 1500 Blue Ridge Road, Box 8616, North Carolina State University, Raleigh, NC 27695–8616)

Schoen, H. (1979). Using the individual interview to assess mathematics learning. Arithmetic Teacher, 27, 34–37.

Simmons, J. (1984). Action research as a means of professionalizing staff development for classroom teachers and school staffs. Grand Rapids, MI: Michigan State University, Teacher Education Center. (ERIC Document Reproduction Service No. ED 275 639)

Vargas, E. M., & Alvarez, H. J. (1992). Mapping out students' abilities. Science Scope, 15, 41–43.

Wood, T., Cobb, P., & Yackel, E. (1991). Change in teaching mathematics: A case study. American Education Research Journal. 28 (3), 587–616.

## Authors Note

**Sarah B. Berenson** is the director of the Center of Research in Mathematics & Science Education and associate professor of mathematics education at North Carolina State University. She has written and directed funded projects totaling $2.5 million. Her honors include the NCSU Alumni Outstanding Extension Award.

**Glenda Carter** is an assistant professor of science education at North Carolina State University. For the past several years her research and public school outreach have focused on authentic assessment. She has been the recipient of federal, state, and local grants to continue work in this area.

# Assessment in Science: A Tool to Transform Teaching and Learning

Margaret A. Jorgensen
James A. Shymansky

There is widespread recognition that schools as they typically are, and teaching as is currently practiced, are failing to prepare all children at high levels. Indeed, some would argue that our educational system is failing to educate all children at even the most basic levels. But what is not characteristic, or even from time to time clear, is what solution course or method should be applied to *fix the problem*.

If we are to move from the rhetoric of education reform into actual practice, the shift in power and responsibility from policymakers to practitioners is a necessity. Inherent in this shift are several tensions. Among the most difficult to resolve:

• Understandings of the roles that teachers *should* play with respect to assessment design, selection, and use—should they be passive implementers of a top-down assessment plan, or should they be active initiators of an assessment plan that supports engaging learning?

• Acceptance of explicit responsibility for student performance—should teachers be responsible for what students learn, or should that responsibility be diluted through the recognition of *uncontrollable* variables

outside the classroom?

• Involvement in a renewal paradigm with ongoing change as the *status quo*—should teachers be expected to advance reform by becoming classroom-based researchers, or should they replicate their successful experiences without revision year after year?

In science education, there is yet another source of tensions—the discipline itself. The debate among the members of the community of scientists and science educators about what must be taught and when it should be taught has contributed to a public sense of fragmentation. This complicates the task for teachers and administrators trying to plan and implement reforms that make a difference in students' understanding, thinking, and use of science.

When these tensions are considered from the perspective of reform in science education, a model for change begins to take shape that is systemic both from the view of the individual teachers and from the view of organized educational practice. Within the model, assessment plays three roles:

• Assessment operationalizes the goals.

Issues in Science Education 107

• Assessment monitors the journey towards those goals.

• Assessment makes explicit the changing roles of teachers from knowledge disseminators to thinking facilitators.

The purpose of this discussion is to present the model with the intent of moving forward in the transformation of teaching and learning in science education towards the powerful proposition that the nation's education reform efforts will accomplish Goal 4 in *America 2000*: "U.S. students will be first in the world in science and mathematics achievement" (1991, p. 3).

## ASSESSMENT TO TRANSFORM TEACHING AND LEARNING

Assessment is at the heart of education reform. It is both the frustration and the hope for lasting change, because it is beyond dispute that *we teach what we test* (e.g., Resnick & Resnick, 1989, pp. 66–67). The frustration stems from a lack of congruence between the education goals for science and the nature of many science tests. The hope stems from the emergence of more *authentic assessments* that document what students think, understand, and can do in science.

Although the word *assessment* has replaced the word *test* and there are now many adjectives juxtaposed to assessment, the essence of reform in testing is that the student is called upon to produce meaningful work. The stimulus for that meaningful work can take many different forms. Often the assignment is to solve a problem, design a model, formulate and deliver a communication, or design an optimal strategy. The work that the student does is theoretically meaningful because it is real—there are explicit, real-world connections between the work the student does and the work in the world beyond school—hence the frequent use of the word *authentic*. However, what must not be forgotten, as assessments depart

more and more from traditional multiple-choice tests to more meaningful tasks, are the fundamental concepts of *systematic administration and systematic scoring*.

In order for equity and fairness of opportunity to endure as assessments change in format and structure, the concepts of systematic administration and systematic scoring are essential. The concepts translate quite simply into practice and do not interfere with the meaningful nature of the work or with the pressing concern to provide optimal opportunity to an increasingly diverse student population. Systematic does not mean rigid or prescribed—it simply means common opportunity and common standards of excellence.

Assessments must be congruent with the content and with the behaviors valued in the course of science instruction. They must also yield information that is meaningful and useful and that informs the teaching and learning process for the microsystem of the individual learner in his/her environment as well as for the macrosystem of groups of learners across diverse environments.

The following figure illustrates the simplicity and dynamics of the proposed model:

By using assessment as the focus for the discussion of lasting change in the way science education is conceived and delivered, educational practitioners can engage in meaningful conversation about the content standards, curriculum, and instruction, and ultimately, about what kind of information meets the needs of schools' customers, i.e., parents, students, teachers, policymakers, and the like.

## GETTING STARTED

A major task confronting educators is how to make sense of the content standards emerging in the disciplines. This task is made more complex in science because of the presence of both the *Benchmarks for Science Literacy* (1993) and the *National Science Education Standards* (1996). It is not the purpose of this article to weigh the relative merits of each or to advise use of one or the other. Indeed, the challenge remains the same regardless of which document is selected to guide the course of science education within a school. That challenge is how to extract from the standards documents those academic expectations that can be appropriately taught and learned during the school experience for all students. Or, as Shavelson (1994) puts it, "deciding what you should teach, even though there are a million things you could teach."

It is inconceivable that all the content of any standards document would define instruction within science education or any other discipline; there is simply too much to cover. If the commitment to depth—the less-is-more adage—also is considered, educators need a strategy or plan to cull through standards documents to shape a teaching-learning program that will contribute to reaching the goal of having U.S. students "first in the world in science ..." (*America 2000*). Assessment conversations can be an effective way to identify from a standards document those academic expectations that are meaningful within the culture of each school while representing national expectations as well.

Teachers must plan an active role in these conversations. In fact, teachers must be the predominant voice. Other members of the school community provide balance as the conversations focus on aspects of science content standards, curriculum decisions, instructional decisions, and information needs. A tried-and-true technique for beginning these conversations, gleaned from parallel experiences in science and other disciplines, is to focus on four specific questions:

- What should be modeled?

- What should be described?

- What should be documented?

- Who should be informed?

Assessment decisions often dictate consideration of the question "Who should be informed?" first. But in this era of education reform, it seems more appropriate to address the question "What should be modeled?" first.

### What Should Be Modeled?

Because teachers do teach what is tested, the science content covered on tests often represents the outline for science instruction. Because teachers do their best to prepare students to recognize and respond to the content on the tests in optimal fashion, the test format also influences instruction. Before the current reform movement began to impact assessment, these realities often led to a quality and content of science education not consistent with what teachers wanted or believed they should be teaching. By changing the nature and substance of assessments, teachers should find compatibility and consistency among assessment, curriculum, and instruction.

So, as teachers think about and discuss what message they want to send to students and to parents through the assessment, they become aware of the power of assessment to transform classroom practice and to empower students as learners. If authentic assessment does little more than focus teachers of science on meaningful performance of tasks within the spirit of scientific inquiry, tremendous reform will have occurred.

Assessment conversations will inevitably lead to questions about the relationship between cooperative groups and *cooperative assessment*. The relationship is volatile; educators generally resist the complications of cooperative assessment—who gets credit for the work if equal effort is not present, and who decides the composition of the groups? Here again the rhetoric of education reform forces consideration of documenting group work in areas of investigation and problem-solving. As the assessment tasks are identified (selected and/or developed), they serve as operational models for the goals of teaching and learning.

## What Should Be Described?

Teachers literally are forced to begin the culling process to address and answer the question "What should be described?" When teachers are asked which three, four, or five *big ideas* in science their students should understand and be able to demonstrate by the end of the school year, the content standards notion quickly becomes, by necessity, a domain of content from which individual teachers must sample. Often the conversations focus on *habits of mind* and on process skills such as systematic inquiry or problem-solving. Sometimes they focus on themes such as the systems, models, change, and energy. At other times they focus on science in terms of real-world tools and applications—technology. These focused conversations rapidly expose and make explicit what classroom teachers think science should be for the students they teach and in the school culture in which they work (Champagne, 1994).

There are not right or wrong directions for these conversations. What is happening is *constructivism in action*, with adult learners making sense of large reference documents that map an entire discipline. Just as young learners construct their own understanding of phenomena and information, so too do adults (Yager, 1991). It is important to recognize this constructivist process in teachers, just as we recognize it and value it in students.

By acknowledging that adults as well as children learn in different ways, with different types of evidence for the acquired knowledge, the tolerance for the products of these conversations expands infinitely. Thus, as teachers are (sometimes) forced into a new responsibility with consequent new power and authority for assessment, they will be supported by the very philosophy that the education reform rhetoric espouses for learners. If the analogy holds, teachers will become active learners, shaping the classroom experience as they come to internalize important science.

## What Should Be Documented?

Having committed to three, four, or five big ideas, the parties to these conversations then need to use a parsimonious attitude to decide what artifacts or documentation of students' thinking, understanding, and doing in science *must* be maintained. In the beginning of the documentation process, there is a tendency to keep literally everything that students do. Whether this is due to a lack of confidence in the selection of evidence or whether it is due to a lack of understanding of space limitations, over the course of a school year it will quickly become apparent that keeping everything just does not make sense.

If one thinks of learning as a journey, it is reasonable to think of assessment tasks as monitoring progress along that journey. At the beginning of the journey, the assessment tasks are probably less complex with respect to the big ideas of science or the habits of mind undergirding the science experiences. Towards the end of the journey through formal education, the assessment tasks are probably more complex as science thinking, understanding, and doing begin to mirror what scientists do in the richest sense.

At points along the school journey, teachers may use the assessment information for a variety of purposes. They can use it to monitor individual student progress and, in a more holistic sense, to monitor the effectiveness of the instructional practices, the curriculum itself, and the congruence between the important science understandings as advocated in content standards documents and those explicit within the school's science goals.

Assessment in support of reform requires the professional consideration of benefit and cost for teachers and students. The introduction of more authentic assessment tasks does not change this basic requirement. Teachers must balance the *cost* of the investment of time and energy that assessment takes against the information derived. Within the rhetoric of reform is a belief that authentic forms of assessment form a seamless web with instruction and learning, thus enhancing the quality of the learning experience for students. It is paramount that teachers strive for this type of experience and that the assessment developed or selected yield meaningful information about important markers along the journey.

Technology often supports the tendency to keep it all, since now almost any artifact, document, or performance can be digitized and stored in a relatively small space. It is important, however, to resist the tendency to keep everything as evidence of what students think, understand, and can do in science. Rather, it is useful to think of one of the fundamental attributes of traditional multiple-choice testing—that of sampling. Standardized norm-referenced or criterion-referenced tests were and are useful because they elicit responses from students over a range of content and behaviors. And, while we may disagree with the test developers' decisions about that content and those behaviors, the tests are built in such a way as to enable the *score consumer* to draw inferences beyond the test itself based only on the test data. This same notion must extend into more authentic tests. If tests yield information applicable only to the task itself and not to a broader application, including behaviors, these new forms of assessment will have little credibility.

One can think of the assessment task itself as disposable. The task must be only a provocation for students to demonstrate the particular aspects of those standards valued within the school culture. The artifact or documentation left by the student after the task has been completed represents the first-stage evidence of explicitly valued learning targets. The second-stage evidence derives from the scores of that documentation.

## Who Should Be Informed?

This last question often will cause considerable discomfort among teachers. It is a difficult task to make explicit the evidence of what students think, understand, and can do that leads to the representation of summary values of the student work. At this point, a database begins to emerge that is constructed of scores of student work; when taken holistically, the scores provide a picture of the learner or group of learners and enable decision-making to begin.

Moving from student work to summary values is essential to the process of building a systematic matrix of evidence about what the student can do. It is inefficient and perhaps impossible to manage information on groups of students if we do not move to a level of abstraction that scoring provides. Thus, this second level of data completes the matrix of evidence and facilitates schoolwide decision-making about content standards, curriculum, and instruction. It also ultimately shapes the information available and the veracity of using the information for individual and/or group or program decisions.

## WINNERS IN THIS TRANSFORMATION PROCESS

Education reform means change, and change means hard work. Sustained reform means continuing hard work on the part of all players. School environments that nurture reform also nurture individuals who take chances, make mistakes, reflect on those errors, and infuse the future with the knowledge of the past. Teachers and administrators who engage in conversations and build action plans to implement the ideas that emerge from these conversations are the wellspring of sustained reform. But accepting the renewal paradigm of ongoing change is not easy for most. It is uncomfortable for many and intolerable for a few. It truly does require an understanding that the science journey for students and teachers alike is always subject to modification as more is learned about the discipline itself and about engaging and meaningful ways to document that journey.

Using assessment as a lever for change to transform the teaching/learning environment in schools across this country will contribute immeasurably to teachers, students, and society. This is the ultimate win/win situation! Teachers become empowered in that they have a new and stronger voice in connecting that which is taught to that which is tested. Students are empowered in that they simultaneously develop powerful habits of mind and learn useful big ideas that will serve them as citizens. These dispositions will provide a platform for science-oriented careers for individuals who choose that path. But, more importantly, these dispositions will ensure that all students have

the tools of systematic inquiry useful in all walks of life and thereby of benefit to society at large.

## References

American Association for the Advancement of Science, Project 2061. (1993). *Benchmarks for science literacy*. New York: Oxford University Press.

Champagne, A.B. (1994). [Interview]. In *Assessment alternatives: Finding out what students know and can do* [video]. Washington, DC: Council of Chief State School Officers.

National Education Goals Panel. (1995). *Goals report*. Washington, DC: Author.

Resnick, L.B., & Resnick, D.P. (1989). Tests as standards of achievement in schools. In *Proceedings of the 1989 ETS Invitational Conference*, 63–80. Princeton, NJ: Educational Testing Service.

Shavelson, R.J. (1994). [Interview]. In *Assessment alternatives: Finding out what students know and can do* [video]. Washington, DC: Council of Chief State School Officers.

Yager, R.E. (1991, September). The constructivist learning model: Towards real reform in science education. *The Science Teacher, 58*(6), 52–57.

## Author Note

**Dr. Margaret Jorgensen** is a senior examiner for the Educational Testing Service. She has directed development teams for norm-referenced and criterion-referenced testing programs at the local, state, and national levels and, most recently, has conducted research and has led development projects in the area of performance-based and portfolio assessment. She is currently working in the broader area of school reform with specific emphasis on mathematics, science, and reading. Dr. Jorgensen has published widely in the area of traditional and innovative assessment.

**James A. Shymansky** is a professor of science education at the University of Iowa.

Author of numerous journal articles, chapters, and textbooks, he has also written and directed many science education grants. Honors include AETS Outstanding Science Educator, NSTA Gustav Ohaus Award for Innovations in College Science Teaching, and the University of Iowa Excellence in Teaching Award.

# Assessing Habits of Mind Through Performance Based Assessment in Science

V. Daniel Ochs

Present reforms in education are extensive and expansive. Past reform efforts have been unidimensional in their focus on curriculum, assessment, instruction, or school organization. Seldom has a reform effort focused simultaneously on more than one of the above. Never has it focused on all simultaneously. The educational enterprise was never treated as a system for reform purposes. When one subsystem (such as a classroom, school, or curriculum) was changed the subsystem was required to function as a piece unsynchronized with the rest of the system. Because of this piecemeal approach few things have really changed or persisted when change was tried.

The current reform is very different in this respect. Changes are occurring simultaneously in curriculum, testing, instructional practice, school organization, and many other facets that interact with each of these subsystems. The result is likely to be meaningful and lasting change. Indeed, while the model for schooling has remained relatively constant over the past 90 years, we are unlikely to see the same model operating as the dominant model in the future. National, state, and district reforms are revolutionizing the educational milieu. Publishing companies are rushing to capitalize on the changes. Testing companies are expanding and shifting the emphases on national tests. States are creating compacts and alliances, sharing expertise and forging new fronts in all areas. Foundations that fund reform efforts are emphasizing systemic reform. Meanwhile, local schools and districts, besieged by the multifaceted approach to reform, are struggling.

## SCIENCE EDUCATION REFORM

Past reform efforts have focused primarily on curriculum. The 1960s saw the development of "science as inquiry" curricula. Attempts to develop teaching practices that were consistent with the curricula met with limited and uneven success. This should come as no surprise since environmental conditions of training workshops and environmental conditions of schools are very different. Teachers returning from training workshops are initially fired up. As they attempt to implement new ideas, their attempts at change are mitigated and they are finally overwhelmed by the status quo and unyielding conditions of the school workplace. They are frequently heard to mutter something like, "It doesn't work." In fact, what they mean is, "I can't make it work given the conditions under which I work." The present broadbased reform efforts are addressing curriculum, instruction, assessment, learning, decision-making processes, and the working environment. They are more likely to meet with success because of this broad-based approach.

Attempts to create instructional practices that were more consistent with visions of the curriculum developers led to the creation of materials in the 1970s that were described as "teacher proof." Some of these programs, most notably the Intermediate Science Curriculum Study, developed teacher materials that included suggestions for organizing the classroom and for creating a variety of assessment strategies. Though "laboratory practical examinations" have been used for years in college classes and in some high schools, this was probably one of the earliest attempts at developing what has come to be known as performance assessment.

Present reform in science education is highlighted by the *National Science Education Standards* (National Research Council, 1996). The standards address reform in many areas: teacher education, teacher classroom behavior, professional development, content, assessment, program, and systemic items such as environmental constraints and energizers.

Current reform efforts change the goals of science education in several important ways. First, there has been a movement away from treating all students as though they are miniature scientists to treating them as citizens who will participate in important decisions about science or will make important decisions that necessarily require an understanding of important science concepts. Second, reform efforts suggest delimiting the content and focusing more on an in-depth understanding of these concepts. Relating concepts to larger themes that permeate all of science is expected to enhance understanding and usefulness of the concepts. Third, reform efforts are viewing the educational enterprise systemically. The numer-

ous conditions and factors that impinge on schools and teachers and the environment in which they work are considered: the culture of schools and localities, state and national policies and reforms, and the interplay among curriculum, instruction, and assessment to name but a few. Finally, there is greater emphasis on thinking skills that are related to science. Within the context of the overall reform is the vision that the scientifically literate citizen will, through the educational enterprise, develop an awareness and internalization of the values, attitudes, and skills that are so useful to the practicing scientist and believed to be useful to all. These values, attitudes, and skills are collectively referred to as habits of mind.

## DEFINING HABITS OF MIND

Over the course of the past several years educators have begun to redefine and expand what is to be developed in the course of schooling the youth of America. In the sciences this is most evident in *Science for All Americans* (Rutherford and Ahlgren, 1990) and the subsequent *Benchmarks for Science Literacy*, (American Association for the Advancement of Science, 1993) and NSES. The first of these documents describes what all students should know *and be able to do* in science, mathematics, and technology by the time they graduate from high school. *Benchmarks* specifies how students should progress toward science literacy, recommending what they should know *and be able to do* by the time they reach certain grade levels. Part of what is to be known and done is described as habits of mind. These habits of mind include scientific values, attitudes and skills. Values and attitudes include things such as curiosity, honesty, openness, and skepticism. The skills valued by the scientific community include computation; estimation; the use of tools that allow enhanced manipulation of materials and organisms and serve as extensions of our senses; and critical response skills that allow one to identify and critically evaluate assumptions and faulty conclusions and suggest alternative ways of explaining events and data. These same skills allow one to identify trade-offs in the design and decision making processes.

Throughout *Benchmarks* there are emphases on practical problems, connections to the real world of the citizen (i.e. scientifically literate citizen, using skills in new situations, outside as well as inside school), and inferences about a world of the future. These inferences can be extended easily to using the information superhighway. There is greater emphasis on locating, organizing, using, and applying information. Although science facts and terms are necessary in understanding and communicating science, reformers decry the rote memorization of factual information. Many of the values, attitudes, and skills described are what science educators have long espoused implicitly. *Science for All Americans, Benchmarks for Science Literacy,* and the *Standards* make these learner expectations explicit. Putting them into instructional practice will be a challenging endeavor because it will require behavioral and attitudinal changes by educators. But even if implementation of teaching strategies designed to foster habits of mind are successful, a number of questions arise. How will we know that habits of mind are learned and used? How can we know *how well* students know and use these habits of mind? What demonstrations or performances will enable us to infer that students have learned these things well? What assessments can we use to find answers to these questions?

## WHAT IS PERFORMANCE BASED ASSESSMENT?

The phrase "know *and be able to do*" appears throughout present-day literature. It is a subtle but extremely important change in emphasis in curriculum, instruction, and assessment. "Know *and be able to do*" explicitly requires that students do something with the knowledge they acquire. They may be

asked to apply knowledge, draw conclusions, make decisions, select or develop alternative scenarios, or manipulate equipment as part of their performance. Goals related to habits of mind have been long espoused in one form or another by science educators, though implementation results have been mixed and difficult to assess. Performance based assessments are particularly well suited to assess many of these goals long espoused by science educators.

More recently, educators from classroom teachers to test developers in large-scale assessment corporations (like ETS, ACT, Advanced Systems in Measurement and Evaluation, and The Psychological Corporation) have investigated performance based assessment examples, practices, validity, reliability and other attributes of these assessment types and strategies. Performance based assessments are appearing more frequently in nationally normed tests. There are, in general, three types of performance based assessments: constructed response, performance event, and performance task.

## Constructed Response Items

Constructed response items frequently include a scenario or real world problem as part of the prompt to which the student responds. Responses appear as short answers, generally written responses. They take from five to fifteen minutes per response, though some may take longer. Constructed response items may require students to interpret data that is given in tables or graphs; they may require students to draw conclusions, make decisions, or identify or weigh alternatives to a problem; some constructed response items are open-ended in nature. Open-ended items usually have two or more possible alternative responses. Students are then asked to justify their response.

## Performance Event Items

Perhaps no assessment strategy has more appeal to science educators than the use of performance events. As many as half the states are using or exploring use of performance event assessments, and performance event items are appearing more frequently in national tests.

New York has a stations approach to its implementation of performance events. Students are given approximately ten minutes at each station where they are asked to perform some task such as completing an electric circuit or classifying a set of items they are given. Over a one-hour period students rotate among several stations dealing with variety in content and demonstrating use of science skills. The situations the students encounter are much like they are assumed to have encountered in the classroom. Consequently, it is difficult to infer that students have really internalized what they have learned. They may well have merely memorized. Test designers are working to ameliorate this problem.

A dramatically different approach to using performance events occurs in California, Connecticut, and Kentucky where students have as much as one hour to solve a single problem. (See Figures 1 and 2)

Usually there are manipulatives that students use to collect or record data. A scenario is developed to set the stage for the *application* of knowledge. The scenario is often taken from a real-world problem. In the performance event illustrated in Figure 1 elementary grade students are asked to test a substance sample collected from a distant

Figure 1.

### DOWNHILL RACERS
Adapted from Kentucky Performance Tasks, 1991–1992.

### PERFORMANCE TASK STUDENT DIRECTIONS/RESPONSE FORM

**GRADE 12**

**STUDENT NAME:** _____

**SCHOOL NAME:** _____

**GENERAL INSTRUCTIONS:**
Some objects will roll down a hill faster than other objects. You have been provided with a ramp and some materials to test. You will be working with one other student for the first 30 minutes of this task. After you have had 30 minutes to do your experiment, the test administrator will ask you to move apart and to finish the task on your own.

You will follow the instructions on this page and on page 4 first. Do not open this form to pages 2 and 3 until you are told to do so.

**MATERIALS:**
- 1 ramp
- 1 stopwatch
- 1 each of the items shown in the chart below.

| Objects in these 2 columns have the same diameter | | Objects in these 2 columns have the same diameter | |
|---|---|---|---|
| glass sphere | steel sphere | glass sphere | steel sphere |
| solid cylinder | solid cylinder | solid cylinder | solid cylinder |
| Hollow cylinder | Hollow cylinder | Hollow cylinder | Hollow cylinder |
| Objects in this column have the same mass | Objects in this column have the same mass | Objects in this column have the same mass | Objects in this column have the same mass |

planet. They are asked to make three tests: (a) looking at and touching the sample; (b) a water absorption test; and (c) an acid or base test. Students are given additional information about the soil requirements of three vegetable types. They collect and organize data and draw conclusions from their observations about the suitability of the material for growing these vegetables in the substance collected from Planet X.

In the second performance event, illustrated in Figure 2, students are given several items that will roll down a ramp. The items vary in several ways. Students are asked to collect data on the different items and variables that might affect the rolling speed and to draw conclusions about which variables most influence speed down the ramp. They are then asked to put what they have learned to use by designing a new item that they predict will roll faster than any they tested.

Performance events may require some background knowledge or background knowledge may be given to the students, as can be seen in the two problems just illustrated. Sometimes students work in small groups during the performance event while they collect data and discuss its meaning. Students are then separated and asked to respond individually to applications and extensions of the concepts tested. The scenario and applications are usually different from what the student has encountered while studying the content in the classroom. This approach raises some questions about student opportunities to learn and about knowledge transfer. These questions are being researched at the present time and results are appearing more frequently in literature. While reports on opportunity to learn and knowledge transfer are mixed, the performance events approach has promise and is growing in popularity.

## Performance Tasks and Portfolios

Another important trend is the growing use of performance tasks and portfolios. The performance task is long-term in nature. Science investigations by students that occur over a period of time, perhaps weeks or months, are particularly well suited to this kind of assessment. Examples are investigations of the distribution of species, habitat changes, chemical composition of waterways, air pollution, and a wide variety of investigations that might be used in competitions such as science fairs, Junior Science and Humanities Symposia, and theWestinghouse Science Talent Search. The results of such an inquiry then become an entry in the portfolio. Where the performance event and constructed response items offer an on-demand snapshot of what a student knows and can do, the performance task and portfolio offer a more comprehensive view of what a student can do with the knowledge acquired in solving real problems. A record of the student's work can appear as a paper or be reported in some other medium, such as a video or computer disk.

Portfolios began as simply containers for student work samples. They are rapidly be-coming much more sophisticated. Portfolios can contain a variety of student work samples and thus provide insight into modes by which students learn and how they communicate what they have learned. These tangential relationships to Howard Gardner's (1983) multiple intelligences are accelerating interest in this assessment format. Two multistate efforts, the State Collaborative for Assessment and Student Standards (SCASS), and the New Standards Project (NSP), are underway to develop and study uses of the portfolio. A generic portfolio model appears in Figure 3.

The laboratory research entry allows a student to identify and carry out a significant investigation of choice. Conclusions would come from data generated by the student in the investigation. The non-laboratory investigative entry relies more heavily on data generated by others. It is typically an issue the everyday citizen might confront, such as a local pollution problem, a family health-related problem, or an issue that might require a voting decision. The creative entry allows students to communicate about science with others in mediums that have not convention-

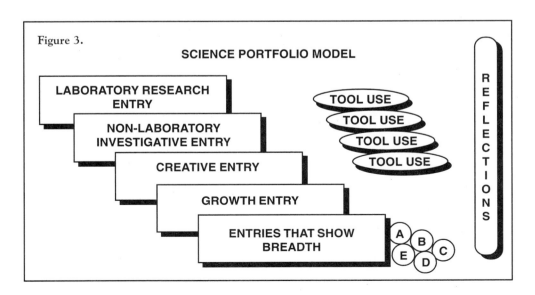

Figure 3.

SCIENCE PORTFOLIO MODEL

LABORATORY RESEARCH ENTRY

NON-LABORATORY INVESTIGATIVE ENTRY

CREATIVE ENTRY

GROWTH ENTRY

ENTRIES THAT SHOW BREADTH

TOOL USE
TOOL USE
TOOL USE
TOOL USE

REFLECTIONS

A B C D E

**Figure 4.**

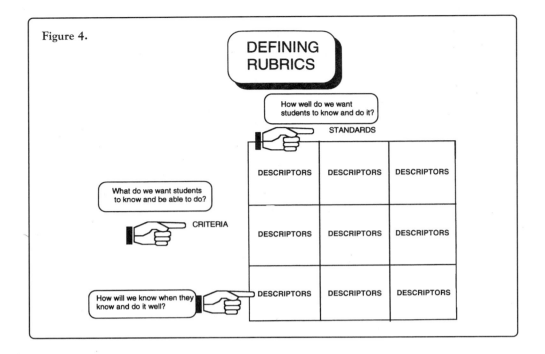

ally been used. Examples are video, art, and music. Such entries would carry strong science story lines, but allow students to communicate what they know using different conventions. A growth entry would show what a student is capable of doing over a period of time, or how a student has changed his or her ideas, or how a concept has been thought about with increasing sophistication. Most of the entries allow students to show depth of understanding. Other entries would show the students' breadth of knowledge in the life, earth, and physical sciences. Students would depict their ability to use tools (e.g. instruments that measure length and volume, instruments such as microscopes that extend the senses in making observations, and instruments such as computer interfaces that enhance the speed with which observations and measurements can be made). Finally, a student would be asked to submit a piece reflecting on each of the other pieces and explaining how the pieces fit together to form a comprehensive picture of what the student knows and can do. The reflections piece

might allow the reader to conclude that the whole picture of the student represented by the entries is greater than the sum of the individual pieces.

Movement toward performance based assessments in states and local districts is growing rapidly. Under the aegis of the Council of Chief State School Officers the State Collaborative for Assessment and Student Standards (SCASS) has evolved. There are presently 15 state members in this collaborative effort. SCASS focuses on the development of a wide variety of alternative assessment strategies including multiple choice, constructed response, performance task, and portfolio. All types of items and the portfolio model have been field tested in participating states. Items are being released for use to the participating states, and development work is continuing. The New Standards Project (NSP), which has sixteen state and six large urban district members, focuses primarily on developing a national model for portfolio assessment. To this end

National Science Teachers Association

the project is identifying science *performance* standards and developing a portfolio model, both of which are compatible with the national content and assessment standards.

## SCORING PERFORMANCE BASED ASSESSMENTS

Scoring of performance based assessments is much different than more traditional forms of assessments. Scoring guides may be designed to score the response holistically or analytically. Holistic scoring guides are usually developed on a four-point scale. Scoring criteria (rubrics) are developed to answer three questions: (a) What do we want students to know and be able to do? (b) How well do we want students to know and be able to do it? (c) How will teachers and other scorers know when the student knows it and does it well? The interaction of the three questions is shown in Figure 4.

Criteria that define the important elements to be scored answer the question of what do we want students to know and be able to do? Descriptors define each level of student accomplishment on the scale and serve to answer the two remaining questions: How well do we want students to know and be able to do these things? And how will teachers and other scorers know when the student knows and does it well?

An analytic scoring guide usually consists of a checklist of points to be covered. Final scoring accounts for how many of the points were addressed in the response.

If rubrics are well done and made available to students, they can serve as a guide to learning. Students can look at the rubrics and know how well they have done and how they can improve.

## USING PERFORMANCE BASED ASSESSMENT TO ASSESS HABITS OF MIND

Habits of mind are difficult to assess us-ing traditional methods of assessment. While multiple choice items can be enhanced to assess higher level thinking skills, many of the habits of mind are more difficult to assess. Effective assessment closely matches the task with the intended student expectation. It sends a message to both teacher and student about what is important, how what is important is to be used, and how we can communicate what we know. The message associated with the learning may be as important as the learning itself. During the 1960s vast amounts of money were spent developing new curricula, and they were good curricula. The vision of the developers was not always implemented in the classroom. True, the environmental factors described earlier contributed to the lack of implementation, but the vision of what was to be done in the classroom and how it was to be done was probably not clearly understood by those who were ultimately responsible for implementation. How else can we explain the deterioration of science as inquiry into a memorization of steps of the scientific method? Attitudes, values, and skills are not adequately assessed on multiple choice items. More importantly, the message we send when we assess these important traits using multiple choice items is clouded by the medium we use to assess students. Important concepts, skills of the scientist, and habits of mind are reduced to memorization lists. Habits of mind typically are represented by how a person acts. We then infer from the action, demonstration, or skill that the person has mastered or internalized the attitude, value, or skill. Performance based assessments are uniquely suited to assessing the habits of mind, because they require students to *perform or demonstrate* something. Students can be asked to collect data through a variety of techniques and using a variety of equipment to extend their senses. They can be asked to organize data they collect in a variety of ways. They can be asked to summarize data, draw conclusions, or make decisions. They can be asked to provide a variety of exhibitions that not only demonstrate what they know and can do, but enable them to respond using many modes of communication.

## PERFORMANCE BASED ASSESSMENT IN THE CLASSROOM

Performance based assessments are well suited to classroom uses. The variety of assessment strategies offers a more complete picture of what a student knows *and can do*. While traditional forms of assessment are still useful, performance based assessments are especially helpful in monitoring habits of mind where the more traditional forms of assessments may be limited. Performance based assessments send a clear message to students and serve as a reminder to teachers that inquiry and habits of mind are important goals of science. These assessments also allow teachers to monitor progress of students through complex and long tasks. If the monitoring is used as feedback, the teacher can provide interventions when students show through their performances that they misunderstand or are not attaining goals of the course, unit, or assignment.

Making the scoring criteria or rubrics public helps students define important ideas, skills, and activities. The development of rubrics as a classroom activity serves to highlight and clarify important aspects of assignments and complex tasks and allows for student "buy-in."

In summary, performance based assessments add an important tool to science educators' ability to communicate and assess what we want known, how well we want it done, and how we can identify when students know and do it well.

## References

American Association for the Advancement of Science, Project 2061. (1993) . *Benchmarks for science literacy*. New York: Oxford University Press.

California Department of Education. (1994, January). *California learning assessment system: A sampler of science assessment, elementary, preliminary edition*. Sacramento, California: Author.

Gardner, H. (1983). *Frames of mind*. New York: Basic Books.

Kentucky Department of Education. (1992). *KIRIS 1991–1992 performance task scoring guide*. Frankfort, KY: Author.

National Research Council. (1996). *National science education standards*. Washington, DC: National Academy Press.

Rutherford, F.J., and Ahlgren, A. (1990). *Science for all Americans*. New York: Oxford University Press.

## Suggested Readings

Herman, J.L., Aschbacher, P., & Winters, L. (1992). *A practical guide to alternative assessment*. Alexandria, VA: Association for Supervision and Curriculum Development.

Marzano, R.J., Pickering, D., and McTighe, J. (1994). *Assessing student outcomes*. Alexandria, VA: Association for Supervision and Curriculum Development.

Perrone, V. (Ed.) (1991). *Expanding student assessment*. Alexandria, Virginia: Association for Supervision and Curriculum Development.

## Author Note

**V. Daniel Ochs** is professor of science education at the University of Louisville. Since 1993 he has been on loan to the Kentucky Department of Education to aid with school restructuring, staff development, and assessment development. He has authored four books and over forty funded proposals. Presently, he directs the science initiative in Kentucky's SSI.

# Science Portfolios:
# Navigating Uncharted Waters

Tom Oppewal

For more than a decade educational discourse has called for a move away from traditional standardized tests to alternative forms of assessment. The widespread dissatisfaction with the *status quo* has produced a deluge of new and not-so-new methods for assessing student learning. These less traditional methods carry names like portfolio, active, alternative, performance, holistic, and authentic assessment. Some challenge established thinking about evaluation and measurement theory. But rather than exploring the spectrum of assessment methods, this article focuses on portfolios and their potential use in science education.

A consensus about what constitutes a portfolio does not exist. The word *portfolio* in educational literature is commonly used to describe "any collection of items." Literally, a portfolio is a container that holds documents that can provide evidence of a person's knowledge and skills (Collins, 1992b; Hamm & Adams, 1991). Such a broad definition is not very useful, however, in helping teachers or policymakers determine a portfolio's purpose or even how portfolios will be developed and assessed in science or other subject areas (Collins, 1992b). Many educators agree that the definition of a

portfolio depends on the intent of the designer and how the portfolio will be used. Hein & Price (1994) suggest the selection of instruments can be "an individual matter, reflecting what a teacher has actually done with the students, who the students are, how old they are, the amount of time that can be given to the assessment, and the purpose of the assessment" (p. 13). The common consensus is that portfolios should not be an expandable file but rather a selection of best works that paint a reasonable picture of a student's learning over time.

Portfolios and other performance-based assessment methods are thought to be more useful than traditional testing for examining students' conceptual understanding and use of complex skills in real-world contexts. Portfolios are also advocated because they allow students to take more ownership and responsibility for their own learning by enabling them to chart their progress and development. Through reflection, students are empowered to become more aware of the processes involved in their own learning.

Although portfolio use is relatively new in many elementary and secondary schools, a long tradition of documenting accomplishments with portfolios exists in the creative and performing arts. Portfolios are also quite popular in the area of writing. There seems to be a natural fit between portfolios and the writing process—especially when students work on multiple drafts and provide input on selecting their best work. The few studies documenting the use of portfolios in science have made it clear that science teachers collect and use some of the same types of student work advocated by experts in portfolio assessment. Student work such as lab reports and research papers have traditionally factored into teacher assessments of student performance. But collecting work from students does not constitute a science portfolio; there should be specific criteria for selecting portfolio materials.

## CONTENT AND ASSESSMENT

What should science portfolios contain? According to Collins (1990; 1992a) portfolios should have a purpose, context, design, and mechanism to score them fairly. Teachers using portfolios in science report including projects, videotapes, drawings, pictures, research reports, and lab reports. Others have suggested that a portfolio can contain material obtained from other types of alternative assessments such as concept maps, interviews, dialogue journals, cooperative group assessments, and open-ended and justified multiple-choice questions (Comfort, 1994; Tippins & Dana, 1992; Tolman, Baird, & Hardy, 1994). The collection of these materials over a period of a year or more enables all stakeholders in the assessment to get a unique picture of the strengths and weaknesses of the student.

A question to consider: What is it in the nature of science learning that makes portfolios an appropriate or inappropriate method of student assessment? Portfolios hold promise for documenting student ability to use critical and creative thinking to solve problems, and also for documenting progress in moving from alternative conceptions toward scientifically acceptable expla-

nations about natural phenomena. However, though many things asked of science students require writing, other non-writing science objectives must also be assessed. Student development in the use of lab equipment, hands-on investigations, social skills, attitudes, and factual knowledge may not lend themselves to portfolio assessment. Also, portfolios in science education can only be used if enough time is given to teaching and learning science. In elementary school classrooms where little time is spent learning science, the portfolio will most likely not be the assessment of choice.

A portfolio that comprises all or a portion of a student's assessment for the school year will need thoughtful decisions about how to make judgments about the quality of student work. These judgments are typically based on a set of criteria—referred to as rubrics. A decision will also need to be made as to whether the portfolio is assigned one score as an entire summary of work or whether a more analytical approach using a series of subscales to separately assess each piece of work is preferable. Ratings are often numerical, indicating achievement on three to six levels. Each level is narrowly defined by level of performance.

While these portfolio assessment methods have found success in some classrooms, portfolios and other alternative assessments become more difficult to use at district and state levels where issues of context, validity, and reliability become increasingly important. Although K–12 students are involved with developing their portfolios, teachers, administrators, and possibly state officials define the purpose of the portfolio and determine how it is used in assessing learning. Teachers need to understand the subject matter, the students, and the stakeholders in order to facilitate successful implementation of science portfolios for student assessment. Science portfolios will need to address a fairly narrow set of standards if they are to be used successfully—national or state science standards are potential starting points for those designing portfolios.

The use of portfolios and other non-traditional forms of assessment is based on a number of assertions. Given that science portfolios are in their infancy, it is prudent to evaluate their use in light of these assertions.

## SALIENT ASSERTIONS RELATED TO PORTFOLIOS

1. *If we change the ways in which students are assessed, it will lead to an improvement in the way teachers teach and conceptualize the curriculum* (Herman & Winters, 1994; Wiggins, 1989). This view is based on the idea that if assessment drives instruction, changing the assessment will alter instruction. This in turn should bring about such student outcomes as the development of higher-level thinking skills, problem solving ability, and better communication in science and other subjects (Guskey, 1994; Paulu, 1994; Sivertsen, 1993).

However, current research findings fail to support these assertions consistently. Some researchers found positive teacher and administrator attitudes toward the use of portfolios. Other studies do not reveal that teachers who use such alternative assessments as portfolios significantly change the way they teach. Change is likely to take place only if high quality professional development opportunities, time, and ongoing support are provided for teachers (Guskey, 1994; Herman & Winters, 1994; Shavelson & Baxter, 1992).

2. *Alternative assessments such as portfolios will reflect real student learning* (Hamm & Adams, 1991; Herman & Winters, 1994; Krest, 1990; Wolf, 1989). There is a growing consensus that traditional tests are not sufficient to assess or to communicate what students have learned. Alternative assessments that relate to learning activities and to real-world situations have persuasive popular appeal. Use of portfolios and other alternative assessments is challenging, however. Student collaboration with peers, teachers, and parents raises important questions as to whether the assessment of a portfolio containing collaborative work is a valid assessment of an individual student's performance (Herman & Winters, 1994; Kagan, 1995). Results of studies comparing how students perform on traditional tests and on portfolio assessment are mixed (Hearne & Schuman, 1992; Herman & Winters, 1994; Shavelson, Baxter, & Pine, 1992).

3. *Using portfolios and other authentic assessments will create greater equity in assessing students.* Traditional assessments do not create an equal opportunity for students with different experiential or cultural backgrounds. Because standardized tests tend to be cross-sectional in time, out of context, and on demand, only the most motivated, test-wise students do well. Portfolios can promote equity by allowing students to compete with themselves rather than others in the district or state (Tippins & Dana, 1992). Portfolio designs can be adapted to varying abilities of students and classes over a long period of time (Krest, 1990). Assessment of a student portfolio must take into account the context and the background of those creating the portfolio. A student's school, classroom, teacher, home environment, and resources available are not equivalent. The desire to use portfolios to make judgments about a student within a district or state must address these and other equity issues when comparing student portfolio scores (Darling-Hammond, 1994; Herman & Winters, 1994; Viadero, 1995). Using a portfolio in high-stakes testing will most likely not enhance efforts to promote equity unless efforts are made to understand the context in which the assessment was given and the unique ways in which knowledge can be expressed.

4. *Teachers, schools, districts, and states are able to obtain a more realistic picture of the impact of resources and the curriculum on student achievement with the use of portfolios* (Haertel, 1990; Hein & Price, 1994; Mills, 1989). Teachers tend to find great value in using portfolios for student assessment. States and districts that have used portfolios to make judgments about student achievement have not found them as useful (Case, 1994). To some degree, this finding relates to pushing the margins of traditional measurement theory. The scoring issue that involves validity and reliability in assessing portfolios continues to vex portfolio advocates (Herman & Winters, 1994; Herman, Gearhart, & Baker, 1993; Wolf, 1993). In addition, time, commitment, resources, and technology are thought to be absent in high-stakes, large-scale assessment (Camp, 1993). Problems described in Great Britain of negative outcomes using writing portfolios are important to consider when thinking about large-scale, high-stakes science assessment (Freedman, 1995). Pioneering work on alternative assessments conducted in states such as Vermont, Connecticut, California, and Kentucky serve as important models. Leaders in these states are wrestling with how best to assess science, despite budgetary and political pressures to return to traditional testing.

5. *While traditional testing has a focus on outcomes, alternative assessment will capture the complex development of student skills, ideas, and concepts* (Kulm & Stuessy, 1991). The portfolio lends itself well to documenting growth in student performance. The information gained from student portfolios is more useful for communicating with students and parents and, compared with standardized testing, assists teachers to help students develop scientific thinking more fully (Camp, 1993). However, it is a challenge to make the switch to portfolio assessment and to communicate with policymakers and parents how to interpret these forms of assess-

ment. Recently policymakers in California and Rhode Island have slowed efforts in using alternative assessments in response to public concern about the usefulness of such reforms. It is clear that the public must be involved from the beginning with any major shift in what students are expected to know and how they are expected to demonstrate that knowledge and skill (Olson, 1995). Maintaining familiar standardized test scores and letter grades on report cards is a strong emotional issue for many parents within the country. A blend of both traditional and alternative assessments is a likely future direction for school districts.

6. *Portfolios and other authentic assessment can bridge the apparent divisions among teaching, learning, and assessment so that they are all integral parts of the process* (Baron, 1990; Comfort, 1994; Reichel, 1994; Wiggins, 1992). The impact that standardized tests have had on teaching and learning is well documented (Herman & Galan, 1993). To counter the effects of teaching to the test, many educators believe that assessment tasks and learning tasks should become interchangeable. The assessment becomes embedded within instruction, so students may not make distinctions between learning and assessment. This may make good sense in theory, but teachers need to teach and organize their classrooms in ways that promote quality science learning. Current studies have indicated that less than half of the fourth grade students in this country are in schools where science is taught on a regular basis. Even science teachers in grades four through six rely on textbooks rather than hands-on experiences (Weiss, 1993). Teachers teaching grades K–12 reported that only one quarter of science class time was devoted to hands-on manipulative activities.

If teachers are to use alternative assessments like portfolios, many will need to first alter their current teaching approaches. As Collins (1992a) points out, because of the meager tradition of formal use of portfolios in science, educators have the potentially exciting challenge of developing a tradition of science portfolios that reflects current thinking in science teaching and learning.

## CHARTING THE COURSE

No assessment method is a panacea. Most would agree that teachers and districts should use a mix of assessment methods to effectively assess student learning. Portfolios have both compelling value and perplexing problems. The popularity that portfolios enjoy presents drawbacks. Literature discussing alternative assessment identifies portfolios on a menu of suggested methods. Portfolios have even been suggested by authors of national reports from governmental and non-government task forces (NRC, 1996; Paulu, 1994). Although portfolios hold great promise, more research is needed to provide directions for those seeking to make decisions about using science portfolios.

Teachers as well as school districts must consider the commitment required for charting a course toward portfolios in science. Time and sustained professional development present challenging hurdles for those seeking to use portfolios or some other alternative assessment methods. This is not just the time it takes to implement portfolios in the classroom but the time it takes teachers to learn how to use and score portfolios (Herman and Winters, 1994). If new assessments are attempted in a school or district, resources will need to be made available. Currently, only a small percentage (16-18 percent) of math and science teachers report having time during the school week to work with peers on curriculum and instruction (Weiss, 1993).

In summary, the potential advantages of portfolios must be tempered with the level of commitment and the available resources. Through creative experimentation by class-room teachers and educational researchers, a new tradition for science portfolios can be forged.

## References

Baron, J.B. (1990). Performance assessment: Blurring the edges among assessment, curriculum and instruction. In *Assessment in the service of instruction*, by A.B.Champagne, B.E. Lovitts, & B.J. Calinger (Eds.), pp. 127–148. Washington, DC: American Association for the Advancement of Science.

Camp, R. (1993). The place of portfolios in our changing views of writing assessment. In *Construction versus choice in cognitive measurement: Issues in constructed response, performance testing, and portfolio assessment*, by R.E. Bennett & W.C. Ward (Eds.), pp. 183–212. Hillsdale, NJ: Erlbaum.

Case, S.H. (1994). Will mandating portfolios undermine their value? *Educational Leadership*, 52(2), 46–47.

Collins, A. (1990). Portfolios for assessing student learning in science: A new name for a familiar idea? In *Assessment in the service of instruction*, by A.B. Champagne, B.E. Lovitts, & B.J. Calinger (Eds.), pp.157–166. Washington, DC: American Association for the Advancement of Science.

Collins, A. (1992a). Portfolios for science education: Issues in purpose, structure, and authenticity. *Science Education*, 76, 451–463.

Collins, A. (1992b). Portfolios: Questions for design. *Science Scope*, 15, 25–27.

Comfort, K.B. (1994). Authentic assessment: A systemic approach in California. *Science & Children*, 32(2), 42–43, 65–66.

Darling-Hammond, L. (1994). Performance-based assessment and educational equity. *Harvard Educational Review*, 64(1), 5–30.

Freedman, S.W. (1995). Exam-based reform stifles student writing in the U.K. *Educational Leadership*, 52(6), 26–29.

Guskey, T.R. (1994). What you assess may *not* be what you get. *Educational Leadership, 51*(6), 51–54.

Haertel, E.H. (1990). Form and function in assessing science education. In *Assessment in the service of instruction,* by A.B. Champagne, B.E. Lovitts, & B.J. Calinger (Eds.), pp. 15–28. Washington, DC: American Association for the Advancement of Science.

Hamm, J., & Adams, D. (1991). Portfolio assessment: It's not just for artists anymore. *The Science Teacher, 58*(5), 18–20.

Hearne, J., & Schuman, S. (1992). *Portfolio assessment: Implementation and use at the elementary level* (Tech. Rep. No. 143). Washington, DC: U.S. Department of Education, Office of Educational Research and Improvement. (ERIC Document Reproduction Service No. ED 349 330).

Hein, G.E., & Price, S. (1994). *Active assessment for active science: A guide for elementary school teachers.* Portsmouth, NH: Heinemann.

Herman, J.L., & Winters, L. (1994). Portfolio research: A slim collection. *Educational Leadership, 52*(2), 48–55.

Herman, J.L., Gearhart, M., & Baker, E.L. (1993). Assessing writing portfolios: Issues in the validity and meaning of scores. *Educational Assessment, 1*(3), 201–224.

Herman, J.L., & Galan, S. (1993). The effects of standardized testing on teaching and schools. *Educational Measurement: Issues and Practice, 12*(4), 20–25, 41–42.

Kagan, S. (1995). Group grades miss the mark. *Educational Leadership, 52*(8), 68–71.

Krest, M. (1990). Adapting the portfolio to meet student needs. *English Journal, 79*(2), 29–34.

Kulm, G., & Stuessy, C. (1991). Assessment in science and mathematics education reform. In *Science assessment in the service of reform,* by G. Kulm & S.M. Malcolm (Eds.), pp. 71–87. Washington, DC: American Association for the Advancement of Science.

Mills, R.P. (1989). Portfolios capture rich array of student performance. *Student Administrator, 46*(11), 8–11.

National Research Council. (1996). *National science education standards.* Washington, DC: National Academy Press.

Olson, L. (1995, April 26). Calif. bill aims to reconstruct testing system: Measure requires chief to design assessment. *Education Week, 14*(31), pp. 1, 16.

Paulu, N. (1994). *Improving math and science assessment* (Report on the Secretary's Third Conference on Mathematics and Science Education). Washington, DC: U.S. Department of Education, Office of Educational Research and Improvement.

Reichel, A.G. (1994). Performance assessment: Five practical approaches. *Science & Children, 32*(2), 21–25.

Shavelson, R.J., & Baxter, G.P. (1992). What we've learned about assessing hands-on science. *Educational Leadership, 49*(8), 20–25.

Shavelson, R.J., Baxter, G.P., & Pine, J. (1992). Performance assessments: Political rhetoric and measurement reality. *Educational Researcher, 21*(4), 22–27.

Sivertsen, M.L. (1993). *Transforming ideas for teaching and learning science: A guide for elementary science education.* Washington, DC: U.S. Department of Education, Office of Research, Office of Educational Research and Improvement.

Tippins, D.J., & Dana, N.F. (1992). Culturally relevant alternative assessment. *Science Scope, 15,* 50–53.

Tolman, M.N., Baird, J.H., & Hardy, G.R. (1994). Let the tool fit the task. *Science & Children, 32*(2), 44–47.

Viadero, D. (1995, April 5). Even as popularity soars, portfolios encounter roadblocks. *Education Week, 14*(28), p. 8.

Weiss, I. (1993). *A profile of science and mathematics education in the United States: 1993.* Chapel Hill, NC: Horizon Research.

Wiggins, G. (1989). Teaching to the (authentic) test. *Educational Leadership, 46*(7), 41–47.

Wiggins, G. (1992). Creating tests worth taking. *Educational Leadership, 49,* 26–33.

Wolf, D.P. (1989). Portfolio assessment: Sampling student work. *Educational Leadership, 46*(7), 35–39.

Wolf, D.P. (1993). Assessment as an episode of learning. In *Construction versus choice in cognitive measurement: Issues in constructed response, performance testing, and portfolio assessment,* by R.E. Bennett & W.C. Ward (Eds.), (pp. 213–240). Hillsdale, NJ: Erlbaum.

## Author Note

**Tom Oppewal** is an assistant professor in science education at East Tennessee State University. His areas of interest include teacher development, student thinking, and the integrated science curriculum.

Issues in Science Education

129

# The Emerging Role of Teacher Leaders: Teachers Speak

Josephine D. Wallace
Catherine R. Nesbit

Anyone who has waded into the active waters of educational reform knows that the experience is complex and challenging. Part of the challenge is that changes by small degrees most often produce small results. For substantial change to occur, the approach must be broad and long-term. According to Hord and Huling-Austin (1986), and Hall and Hord (1987) the process of educational change is complex and long term. The literature on effective schools notes that teachers and administrators must be involved in the process if school change is to take place (Martin, Green, and Palaich, 1986; Hall & Hord, 1987). With school reform, new teacher leader roles are developing (Lieberman & Miller, 1990). While these teacher leaders are recognized as accomplished classroom teachers, they do not necessarily possess the knowledge and skills to take on new roles as leaders (Manthei, 1992). Consequently, they need opportunities to develop leadership skills necessary to facilitate school improvement. Lieberman, Saxl, and Miles (1988), Devaney (1987), Price (1990), and Gehrke (1991) identify common elements to include in effective teacher leader programs: decision-making, curriculum knowledge, peer teaching and feedback, and the principles of leading and organizing staff development. Others (Sparks, 1983; Harty & Enochs, 1985; Taylor, 1986; National Science Board, 1988; Graebill & Phillips, 1990; Hord & Czerwinkski, 1991; & Loucks-Horsley, 1992) have suggested additional factors to promote effective professional development programs. They include involvement of teachers in the planning and implementing process, programs that directly relate to classroom instruction, articulation of a vision, monitoring, and active administrator involvement. There have been a number of studies on important factors to include in leadership development; these studies have been primarily based on program developers' perspectives and not on the participants' perspectives of these programs (Gehrke, 1991). This article will describe seven different models of lead teacher roles presented to teachers, how one of these roles evolved over time, and the experiences that teachers identified as significant in preparing them for these leadership roles.

## A SOLUTION FOR REFORM

A team of school-based lead teachers working closely with their principal was selected as the vehicle to improve elementary school science and mathematics. The idea was to assist schools with site-based improvement goals by helping the teams of lead teachers facilitate the change process in their schools through a professional development program. Over a three year period, a team of two lead teachers and their principal at each of 180 schools was selected to work with the other teachers at their school to bring about this change. The project, conducted by The University of North Carolina Mathematics and Science Education Network (MSEN), was funded by the National Eisenhower Program Fund for the Improvement and Reform of Schools and Teaching (FIRST). The FIRST project was held at eight MSEN university-based Mathematics and Science Education Centers across the state.

## WHO WERE THE TEACHER LEADERS?

This project referred to teacher leaders as lead teachers. Lead teachers were full-time elementary school teachers who had regular teaching responsibilities and took on the additional leadership role of attempting to strengthen the way all teachers at their schools taught science and/or mathematics. Some of the responsibilities of this position included (a) serving as the resource person for the school in the area of science and/or mathematics by keeping peer teachers informed of current information, resources, and teaching practices; (b) serving as the school contact for science and/or mathematics; (c) facilitating and conducting workshops for other teachers; (d) demonstrating effective classroom instruction; and (e) mentoring and coaching fellow teachers.

## AN OVERVIEW OF THE PROFESSIONAL DEVELOPMENT PROGRAMS

As an initial step in strengthening the way science and/or mathematics were taught, all teachers and administrators at each school completed a MSEN program assessment instrument (Franklin, 1990; Penta, Mitchell, & Franklin, 1993) that assessed the strengths and the weaknesses of their school's science and/or mathematics programs in four major

areas: curriculum, instruction, assessment, and school climate. The instrument was adapted from the National Science Teachers Association (NSTA) School Science Program Guidelines for Self-Assessment (Voss, 1987). The results of each school's assessment were synthesized by the lead teachers. Using the assessment feedback, each team developed a School Improvement Plan (SIP) for meeting its school's needs in science and/or mathematics. The team reassessed and revised the plans as needed. Then program coordinators at each of the eight university sites planned the professional development program for lead teachers that was designed to help each school carry out its plan. These eight different programs were designed to cover the topics identified by the assessment instrument. The topics included a varied mix of activities with a strong focus on content and pedagogy. In addition, each program included a leadership development component. The entire professional development program included three Pre-Assessment Sessions and a 75-contact-hour Summer Institute, followed by six Academic Year Sessions and a 25-contact-hour Final Summer Workshop.

The challenge for the developers was to tap lead teachers' concerns about the delivery of science and/or mathematics instruction, not only in their own classrooms, but also to engender their willingness to take on a leadership role in which they would involve all teachers at their schools in improving the way science and/or mathematics was taught. To develop the capacity to accomplish this task, the lead teachers had to sharpen their leadership skills. The program coordinators knew that declaring a teacher *lead teacher* does not necessarily make it happen. So they set about the task of creating a comprehensive plan that would prepare the lead teachers for this role.

The professional development program designed by the program coordinators at each site varied in the amount of time devoted to

**Table 1.**

### Leadership Development Topics/Activities Presented at Program Sites

| Topic: | Number of Programs |
|---|---|
| Cooperative Learning | 15 |
| Sharing Activities, Resources, Presenting to Others | 13 |
| Processing/Discussing Their Leadership Experiences, Problem Solving | 13 |
| Working with Peers, Team Building, Involving Others, Communicating | 11 |
| Teacher Change, Change Process; CBAM Model | 9 |
| Leadership Styles, Roles, Models | 8 |
| Obtaining Funds | 8 |
| Peer Coaching | 8 |
| Adult Learning/Learning Styles | 7 |
| Workshop Design and Presentation | 7 |
| School Culture | 5 |
| Networking | 4 |
| Managing and Utilizing Resources | 3 |
| Peer Mentoring | 2 |
| Time Management | 1 |
| Professional Activity Involvement | 1 |
| Assessing Leadership Skills | 1 |
| Involving the Principal | 1 |
| Assessing Peer Skills | 1 |

*Note.* From "Statewide improvement in elementary mathematics and science education through peer teacher training," by M. E. Franklin, 1993, *Final Report of Project R186000258–92*, p. 66. Reprinted with permission.

content and pedagogy versus leadership and planning. Time spent on content and pedagogy at each site ranged from 60 percent to 90 percent. Focus on leadership and planning varied from 10 percent of total time to 40 percent. Many sites brought in outside presenters who spoke about leadership topics such as team building, the change process, and peer coaching. Leadership topics most frequently presented during the professional development sequence included (a) cooperative learning; (b) sharing activities, resources, presenting to others; (c) processing their leadership experiences and problem solving; (d) working with peers, team building, involving others, communicating; (e) teacher change, change process, Concerns Based Adoption Model; (f) leadership styles, peer coaching, adult learning; and (g) workshop design and presentation (Table 1). Sites varied from low to high in the amount of time focused on practicing leadership skills such as co-teaching or presenting to their peers. Some strategies expected lead teachers to conduct inservice and use peer coaching. All sites modeled leadership skills by presenters, often by master teachers, and involved the teachers in presenting lessons to their peers during the professional development sequence. Many sites involved lead teachers in leadership role-playing exercises.

Ongoing assessment of lead teachers' concerns was a key element of the process. Throughout the implementation, assessments of lead teachers' needs refined the professional development process. These assessments were carried out through formative evaluation of lead teachers' needs and reactions using periodic administration of the Concerns Based Adoption Model (CBAM) (Hall, George, & Rutherford, 1986; James, Hord, & Pratt, 1988) at all sites and through focus group interviews at several sites. The CBAM helped the developers respond to lead teachers' changing concerns. In addition to these formal evaluation modes, a strategy at one site to collect data informally used a vehicle called *home groups* that allowed time for lead teachers to work in small groups to share their ideas, develop trust, and assess the overall progress of the Institute.

## WHAT ROLES DID THE TEACHER LEADERS TAKE?

As a result of the professional development programs designed by the program coordinators (Nesbit, Wallace, & Miller, 1995), and based on the needs identified by the teachers in the assessment instrument, seven different leadership models (Table 2) emerged at eight program sites. The leadership models as given in Table 2 varied considerably on two dimensions, *proactivity* and *sphere of influence*. Lead teachers demonstrated *proactivity* when they initiated the involvement of other teachers in bringing about school change. Proactivity varied from no proactivity in the Classroom Role Model to a high level of proactivity in the Change Agent Role. *Sphere of influence* related to the area where the change took place. Change could take place in individual classrooms or throughout the school. It shifted from the classroom level with the Classroom Role to

the whole school with the Change Agent Role. The seven models (Nesbit, Wallace, & Miller, 1995, pp. 6–9) are as follows:

1. *Classroom Role Models* exemplify good teaching within the subject area in their classrooms. They are not proactive in going to other teachers but respond by providing support when asked for assistance. They maintain a low profile on the school level, but their influence is felt directly in their own classrooms and indirectly in their peers' classrooms.

2. *Active Classroom Role Models* seek out other teachers to share informally with them ideas they practice in their classrooms. They are somewhat proactive because they offer help to others without being asked.

3. *School Role Models* inform the school faculty about a variety of sources for improving teaching in their classrooms. This may be done at the individual, grade, or school level. They are fairly proactive, because they may influence the whole school with the information they share.

4. *Manager Role Models* directly promote instructional change by ordering and maintaining materials and organizing workshops for school faculty. They are proactive by providing expertise in actively supporting and influencing the subject area on a school-wide basis.

5. *Instructor Role Models* are officially designated by the principal to establish and clarify norms and expectations for effective teaching of the subject. They are very proactive in leading workshops related to subject area and influencing the implementation of administrative directives.

6. *Change Agent Role Models* inspire the faculty to get involved in deciding the changes they collectively wish to implement at the school level. They are highly proactive in

**Table 2.** Program Coordinators' Leadership Models

| MODEL NAME | CLASSROOM ROLE MODEL | ACTIVE CLASSROOM ROLE MODEL | SCHOOL ROLE MODEL |
|---|---|---|---|
| Level of Proactivity | Not Proactive | Somewhat Proactive | Fairly Proactive |
| Sphere of Influence | Classroom | Classroom | Classroom/school |
| DESCRIPTIONS<br><br>Lead teachers in this model… | exemplify good teaching within the subject area in their classroom. They are not proactive in going to other teachers but respond by providing support when asked for assistance. They maintain a low profile on the school level but their influence is felt directly in their own classrooms and indirectly in their peers' classrooms. | seek out other teachers to share informally with them ideas they practice in their classrooms. They are somewhat proactive because they offer help to others without being asked. | inform the school faculty about a variety of sources for improving teaching in their classrooms. This may be done at the individual, grade, or school level. They are fairly proactive because they may influence the whole school with the information they share. |
| CHARACTERISTICS | Is a role model in own classroom<br><br>Indirectly influences others.<br><br>Has subject matter and teaching expertise.<br><br>Is a resource or resident expert who will help if asked. | Offers support actively and is available to one or more teachers on an informal basis.<br><br>Is a role model in own classroom.<br><br>Has subject matter and teaching expertise.<br><br>Is a resource or resident expert to whom teachers come. | Shares willingly at the school-wide level new ideas and resources.<br><br>Communicates informally with small or large groups.<br><br>Is a role model in own classroom.<br><br>Has subject matter and teaching expertise.<br><br>Is a resource or resident expert to whom teachers come. |
| Less Emphasis on: | Maintaining materials<br><br>Challenging, inspiring, motivating.<br><br>Collaborating<br><br>Teaching others formally.<br><br>Developing leadership skills. | Maintaining materials.<br><br>Collaborating<br><br>Challenging, inspiring, motivating.<br><br>Teaching others formally. | Maintaining materials.<br><br>Teaching others formally.<br><br>Challenging, inspiring, motivating.<br><br>Democratic decision making. |

initiating ideas and getting others to generate ideas, participate, and maintain the process necessary to make changes related to the subject.

7. *Variable Role Models* exhibit characteristics from one or more of the models of lead teacher roles. Most teachers demonstrate varying levels of proactivity and spheres of influence based on personal and/or situational factors.

These leadership models were formulated and categorized at the conclusion of the project by the statewide Project Director and three other researchers. The models emerged as a result of observing a portion of the leadership component at each site,

| MANAGER ROLE MODEL | INSTRUCTOR ROLE MODEL | CHANGE AGENT ROLE MODEL | VARIABLE ROLE MODEL |
|---|---|---|---|
| Proactive | Very Proactive | Highly Proactive | Selectively Proactive |
| School | School | School | |
| directly promote instructional change by ordering and maintaining materials and organizing workshops for school faculty. They are proactive by providing expertise in actively supporting and influencing the subject area on a school-wide basis. | are officially designated by the principal to establish and clarify norms and expectations for effective teaching of the subject. They are very proactive in leading workshops related to subject area and influencing the implementation of directives. | inspire the faculty to get involved in deciding the changes they collectively wish to implement at the school level. They are highly proactive in initiating ideas and getting others to generate ideas, participate, and maintain the process necessary to make changes related to subject. | exhibit characteristics from one or more of the models of lead teacher roles. Teachers demonstrate varying levels of proactivity and spheres of influence based on personal and/or situational factors. |
| Organizes materials for school staff in lab or other area. | Is authorized by the administration to give direction to peers about subject area. | Challenges, inspires, motivates peers to initiate school-wide change. | Is a role model in own classroom. |
| Obtains and maintains materials. | Conducts workshops. | Collaborates and enables peers to effect change. | Has subject matter and teaching expertise. |
| Coordinates workshop for school or grade level. | Takes charge of instruction in subject area. | Promotes democratic decision making. | Shares, helps, guides. |
| Solves problems related to subject area. | May observe and evaluate peers in subject implementation. | Develops leadership skills in self and others. | Maintains materials. |
| Is school contact or liaison for subject. | | Has subject matter and teaching expertise. | Challenges, inspires, and motivates. |
| | | | Is a resource or resident expert. |
| Directly promotes and advocates for subject. | | Is a resource or resident expertise. | Collaborates. |
| | | | Develops leadership skills. |
| | | | Democratic decision making. |
| Being a role model in own classroom. | Maintaining materials. | Maintaining materials. | Various characteristics depending on model. |
| Challenging, inspiring, motivating. | Being a role model in own classroom. | Being a role model in own classroom. | |
| Democratic decision making. | Challenging, inspiring, motivating. | | |
| | Collaborating. | | |
| | Democratic decision making. | | |

through telephone interviewing of the program coordinators and their teaching staff, and examining program documents submitted by the program coordinators. Program documents included Institute schedules, syllabi, questionnaires, and reports. The telephone interviews and program documents included questions such as (a) How would you define a lead teacher? (b) What do they do? (c) What is their role?

## TEACHERS' REFLECTIONS ABOUT THEIR EMERGING LEADERSHIP ROLES

Eight different professional development programs were planned and implemented by the project coordinators. One science improvement program designed for the Change

Agent Role provides a window to lead teachers' reflections about their emerging leadership role as they progressed through the program. Focus group interviews gathered these teachers' reflections at four different times during the Change Agent professional development program (Nesbit & Wallace, 1994). This technique gathered testimony on the perception of leadership from six to eight lead teachers at a time as they progressed through the staff development process. During the interviews, they were asked (a) How would you describe your role as lead science teacher? (b) People often have different models when they think about teachers as leaders, what is your model of teacher leadership? (c) When you think about your role as a lead teacher now, is it the same or different from the way you expected it to be at the beginning of the project?

At the beginning of the professional development program for the Change Agent Role and at the first interview, most teachers viewed their role not as a Change Agent but as a *responder*. In other words, they viewed a lead teacher as a person who would help only if teachers came to them and asked. An example would be teachers asking the lead teacher to order or locate science materials and equipment for their use. In addition, the lead teachers viewed their role as *autocratic*, which meant they did not involve other teachers in the decision making process. They believed they, as lead teachers, would get the information necessary to have an effective school science program at the Summer Institute. Then all they had to do was go back to their schools and tell the other teachers what they learned.

By the second interview, many of the lead teachers saw their leadership role in a different light. They began to see themselves more as Change Agents; that is as *initiators* rather than as *responders*. They now saw their roles as more proactive; as one of going to other teachers and initiating action,

not merely responding when teachers came to them for assistance. This shift in perspective was expressed by a teacher in the second interview when she said:

> I originally felt that as the lead teacher I would help teachers only if they came to me for help. Now I think I will be taking a more direct approach and will be encouraging them and going to them to make changes. (Nesbit & Wallace 1994, p.6)

By the second interview the lead teachers were also beginning to see the importance of getting their teachers involved in decision making. In other words, they were beginning to view themselves as working with their peers democratically rather than autocratically. This involvement in decision making showed itself in a variety of ways. For example, many took their School Improvement Plans back to their schools to get teachers to support it. This shift was expressed well by a teacher when she said:

> When I first thought about it [the lead teacher role], I thought that all I'd have to do was come down here [to the Summer Institute] and get the information and just give it to them [the school faculty] and let them do whatever they wanted to. But I see now that I need to get more of my faculty involved. (Nesbit & Wallace, 1994 p. 6–7)

By the third and fourth interviews, many teachers were viewing their Change Agent Role as being more *democratic* and *initiating*. In order to get other teachers involved, the lead teachers often worked one-on-one with their teachers to initiate change. Here is one teacher's story:

> At our school I think one of the things that has changed for me, particularly as a lead teacher, is that now I've become a counselor to the teachers with

experience. Beginning teachers are willing to try anything. So I have to encourage them [the veteran teachers] by almost providing the lesson. I've done that for some teachers who've taught as long as I have. So after my lesson, after a hands-on science lesson, I would deliver it to her [veteran teacher] room and explain to her how I would use these things and once she'd gotten into it, she really enjoyed it. And the children enjoyed it. But I did all the work. So we now have this agreement that she would do the next one and she will do the same thing with me. She will get all the materials ready, and then she will bring it to me when she's finished. So we're sharing all the way. (Nesbit & Wallace, 1994. p.7)

Getting others involved and finding out their ideas were done in a number of ways. Some of the examples ranged from faculty sharing their ideas for hands-on science activities at faculty meetings, to the whole faculty presenting a Family Science Night at a PTA meeting.

Even though all lead teachers at this site received the leadership development program for the Change Agent Role, the focus group interviews showed they did not all respond to the leadership role in exactly the same ways. Instead individual teachers progressed to different points along the continuum of proactivity and shared decision making depending upon their own personal leadership styles and the climates at their schools. However, all lead teachers practiced some degree of proactivity and shared decision making. They all moved toward being more *democratic* and *initiating*, which exemplified characteristics of the Change Agent Leadership Role in that they actively involved their teachers in the process of change.

## WHAT DID TEACHER LEADERS SAY WERE CRITICAL FACTORS?

Lead teachers involved in the eight different professional development programs were asked, "What were the significant experiences in the professional development sequence that prepared you for your role as a lead teacher?" Most responses referred to (a) the hands-on approach to the professional development sequence, (b) the presentation of up-to-date science and mathematics content and current curricular programs, and (c) the opportunity to talk to other teachers about common problems. Other helpful aspects of the professional development sequence cited by lead teachers included the opportunity to visit and observe at other lead teachers' schools and school site visitations by the program teaching staff. These experiences facilitated collaboration and collegial support. An example of collegial support involved teachers working closely together as peer coaches and as a support system for each other. Lead teachers also expressed the importance of the site teaching staff modeling effective teaching strategies and leadership techniques to prepare them for their new leadership roles.

Key topics of the leadership development component included coverage of difficulties that come with the change process and the time it takes to make change. A further critical element noted by the lead teachers was that they needed principals who cooperated with them extensively and provided the administrative support for implementing change in their school.

Primary difficulties encountered by lead teachers included poor working relationships with the principal or between the two lead teachers, and not enough time devoted to leadership development. Other administrative factors that impeded their ability to successfully implement the project included (a) a change in the school principal, (b) the absence of a second lead teacher, and (c) the need for time and guidance from program coordinators and their teaching staff for planning and implementing workshops.

Promising insights were gained from the FIRST project. Teachers can function as leaders in many ways—seven leadership models emerged from the statewide professional development project. Teacher leaders evolved in their roles over time. Teachers identified important experiences that prepared them for their leadership roles. Teacher leaders in this project have made new inroads in school-based science and/or mathematics reform by moving beyond the status quo, influencing their colleagues' thinking and practice, and thus improving science and/or mathematics instruction in their schools.

## References

Devaney, K., (1987). *The lead teacher: Ways to begin*. New York: Carnegie Forum on Education and the Economy.

Franklin, M.E. (1990). *Elementary school science and mathematics program assessment*. Chapel Hill, NC: The University of North Carolina Mathematics and Science Education Network.

Franklin, M.E. (1993). *Statewide improvement in elementary mathematics and science education through peer teaching training*. (Final Report of Project R168D00258-92). Washington, DC: U. S. Department of Education.

Graebill, L., & Phillips, E. (1990). A summer math institute for elementary teachers: Development implementation, and follow-up. *School Science and Mathematics*, 2, 134–141.

Gehrke, N. (1991). Developing teachers' leadership skills. *ERIC Digest*. Washington, DC: Office of Educational Research and Improvement.

Hall, G.E., & Hord, S.M. (1987). *Change in schools: Facilitating the process*. Albany, NY: State University of New York Press.

Hall, G.E., George, A.A., & Rutherford, W.L. (1986). *Measuring stages of concern about the innovation: A manual for use of the SoC questionnaire*. Austin, TX: The University of Texas at Austin, Research and Development Center for Teacher Education. (ERIC Document Reproduction Services No. ED 147 342) Distributed by Southwest Educational Development Laboratory, Austin, TX: (Original work published in 1979 as R & D Report No. 3032)

Harty, H., & Enochs, L. (1985). Toward reshaping the inservice education of science teachers. *School Science and Mathematics*, 2, 125–135.

Hord, S.M., & Czerwinski, P. (1991). Leadership: An imperative for successful change. *Issues about Change*, 1(2), 1–6.

Hord, S.M., & Huling-Austin, L. (1986). Effective curriculum implementation: Some promising new insights. *Elementary School Journal*, 87(1), 97–115.

James, R.K., Hord, S. M., & Pratt, H. (1988). Managing change in the science program. In *Third Sourcebook for Science Supervisors*, by L. Motz & G. Madrazo (Eds.). Washington, DC: National Science Teachers Association, 61–75.

Lieberman, A., & Miller, L. (1990). Teacher development in professional practice schools. *Teachers College Record*, 92(1), 105–122.

Lieberman, A., Saxl, E., & Miles, M. (1988). Teacher leadership: Ideology and practice. In *Building a Professional Culture in Schools*, by A. Lieberman (Ed.). New York: Teachers College Press.

Loucks-Horsley, L. (1992). *Effective teacher development programs*. Presentation of the National Eisenhower Conference, Washington, DC.

Manthei, J. (1992, April). *The mentor teacher as leader: The motives, characteristics and needs of seventy-three experienced teachers who seek a new leadership role*. Paper presented at the annual meeting of the American Educational Research Association, San Francisco, CA.

Martin, J., Green, J., & Palaich, R. (1986). Making teachers partners in reform. *Teaching in America: The possible renaissance*. Denver, CO: Education Commission of the States.

National Science Board. (1988). *Science and engineering indicators—1987*. Washington, DC: National Science Foundation.

Nesbit, C.R., & Wallace, J.D. (1994, March). *The impact of leadership development on perceptions of the elementary lead science teacher role*. Paper presented at the annual meeting of the National Association for Research in Science Teaching, Anaheim, CA.

Nesbit, C.R., Wallace, J.D., & Newman, C. (1993, Fall). Building a school's science leadership team. *Journal of the North Carolina Science Teachers Association*, 19–23.

Nesbit, C.R., Wallace, J.D., & Miller, A-C. (1995, April). *A comparison of program co-Coordinators' leadership models and lead teachers' implementation of those models in the schools: Is there a match?* Paper presented at the annual meeting of the National Association for Research in Science Teaching, San Francisco, CA.

Penta, M., Mitchell, G., & Franklin, M. (1993). *Reliability studies of a needs assessment instrument for elementary school mathematics and science programs in North Carolina*. Paper presented at the Annual Meeting of the North Carolina Association for Research in Education, Greensboro, NC.

Price, E.C. (1990, August). *Enhancing the professionalization of teachers through effective leadership training*. Paper presented at the summer workshop of the Association of Teacher Educators, Baltimore, MD.

Sparks, G.M. (1983). Synthesis of research on staff development for effective teaching. *Educational Leadership, 52*, 65–72.

Taylor, R. (1986). *Professional development for teachers of mathematics: A handbook*. Reston, VA: National Council of Teachers of Mathematics.

Voss, B. (1987). *Guidelines for self-assessment for elementary school programs*. Washington, DC: National Science Teachers Association.

## Author Note

**Catherine R. Nesbit** is a science educator and associate professor in the Department of Curriculum and Instruction at the University of North Carolina at Charlotte. She has worked extensively with teachers on cooperative learning, constructivist learning, and gender equity in science in the United States and Africa. She has been involved in several curriculum development projects which resulted in the production of an activity book for teachers, Science EQUALS Success, and a sourcebook which integrates African cultures into the curriculum. Her present research interest is teacher leadership development.

**Josephine D. Wallace** is director of the Mathematics and Science Center and assistant professor in the Department of Curriculum and Instruction at the University of North Carolina at Charlotte. She teaches undergraduate and graduate courses in science education and has received grants on the state and national levels to support the professional development of teachers in science. She is the author of articles in publications of the National Science Teachers Association and the National Association for Research in Science Teaching. Her research interests include examining student conceptions in science through the use of concept mapping and teacher leadership development in science.

# The Constructivist Leader

M. Gail Jones

"Constructivism reminds us that order exists in the minds of people; so when we as [leaders] impose our order on [our colleagues], we rob them of the opportunity to create knowledge and understanding themselves" (Brooks, 1990, p. 70).

## CONSTRUCTIVISM

One of the most intriguing reform efforts taking place in science education today is the recent move toward constructivist teaching. Understanding the theoretical foundations of constructivism is essential for leaders in science education making curricular and policy-related decisions.

Constructivism is based on the premise that knowledge is not something that can be transferred from one person to another, but instead must be built by the individual. As a result, learning is a highly idiosyncratic process. According to constructivist theory it is not possible for one person to pass knowledge on to another person. As a consequence, the teacher's role is that of a facilitator, an engineer of the learning environment assisting students in developing understandings of science phenomena.

Radical constructivists take constructivism to a more extreme position and maintain that we cannot directly understand reality or objective truth that exists outside of ourselves. Radical constructivists believe that our knowledge of the world is a result of our experiences. "Constructivism replaces the notion of truth with that of viability, which does not refer to anything outside of the experiential field" (von Glasersfeld, 1993, p. 27). The concept of viability is a critical component of constructivist theory. Individuals seek viability of ideas as they build understandings of experiences and attempt to fit these understandings with ideas held by others.

The increasing interest in constructivism appears to have arisen from a dissatisfaction with the extreme application of behaviorist theory in American schools. During the last twenty years the educational system has been driven by calls for instruction by objective, behavioral contracts, mastery learning, and stimulus-response types of lessons. Traditionally the teacher's role has been one of an actor who must make all the right moves to get students to master knowledge. Testing programs have also mirrored traditional behaviorist models of instruction by providing strict objective tests to determine the extent to which the material was mastered or absorbed. Science educators have finally spoken out against this type of positivist model and have shifted the focus from the teacher to the students, recognizing that learning is the product of student experiences.

## SOCIAL CONSTRUCTIVISM

One of the most useful ways of making sense of the world is through interactions with others. Social constructivists argue that individuals find meaning not only through individual experiences but also through social interactions. A critical component of social constructivism is the emphasis on the role of language in learning. Language enables us to think about our thinking as well as to go beyond simple stimulus-response modes of thinking to higher forms of critical thinking skills. Language also assists the learner in determining the viability of ideas. Social constructivism is based on beliefs that knowledge is embedded in culture and is shared with others through written and verbal language. Science, to a social constructivist, is not objective, unbiased truth waiting to be discovered, but rather is an intellectual construct that results from human activity (Nadeau & Desautels, 1984). The scientific process includes not only the processes of asking questions, collecting and analyzing data, but also includes communicating results to others for replication and validation. Thus, science is a social process.

Learning science, according to Driver, Asoko, Leach, Mortimer, and Scott (1994):

Involves being initiated into scientific ways of knowing. Science entities and ideas, which are constructed, validated, and communicated through the cultural institutions of science, are unlikely to be discovered by their own empirical enquiry; learning science thus involves being initiated into the ideas and practices of the scientific community and making these ideas and practices meaningful at the individual level. The role of the science educator is to mediate scientific knowledge for learners, to help them to make personal sense of the ways in which knowledge claims are generalized and validated. (p. 6)

## THE CONSTRUCTIVIST TEACHER

Constructivist philosophy alters the fundamental ways we view the teaching-learning process. The focus of a constructivist science classroom is on *construction of meaning*. This collaborative process involves the teacher, the student, and peers. The role of the science teacher is to facilitate and mediate the construction of knowledge. This primary role changes the traditional nature of the teaching process. It is not possible to "cover a curriculum" or "get through a textbook." These metaphors imply a learning process in which the teacher takes action that automatically results in "achievement." Instead, science classes are places for exploration, discovery, and building understandings. Laboratories no longer serve only to verify the contents of a lecture, but instead serve as rich environments where students develop ideas, experiments, and models that become springboards for new ideas, experiments, and models. The constructivist teacher uses inquiry models such as the learning cycle to promote cognitive engagement and growth in students.

The learning cycle (Lawson, Abraham, & Renner, 1989) is designed to promote disequilibrium, argumentation, and improved reasoning. A learning cycle lesson includes three phases—exploration, term introduction, and application (See Table 1). A typical learning cycle lesson begins with exploration, in which the teacher attempts to find out what ideas students hold. This social process allows students to discuss and discover conflicts between their ideas and those of others. During exploration students ask questions, develop hypotheses, and make predictions. The collection and analysis of data further encourages students to refine their ideas and generate new questions and hypotheses. This process, conducted with others, promotes higher order thinking skills. During the second phase of the learning cycle, students report their data, describe patterns, and share terminology and other information gained from references. In the final phase of the learning cycle students explore applications that involve the same phenomena. This links the new understandings with relevant experiences and assists the student in making connections with prior knowledge.

## CONSTRUCTIVIST ASSESSMENT

If the goal of instruction is to assist students in building their own understandings of phenomena, then traditional paper-and-pencil multiple-choice tests may not be the

---

**Table 1.** *Using The Learning Cycle To Teach for Understanding*

**EXPLORATION**

Students observe and describe phenomena.

Students reexamine phenomena.

Different student perceptions are shared and discussed.

Teachers and students ask numerous questions.

Student responses are listened to very carefully—responses are probed for clarification.

Students develop hypotheses, identify variables, collect and analyze data.

Students draw initial conclusions about phenomena.

**TERM INTRODUCTION**

Students compare data.

Terms are introduced.

Students design models to explain phenomena.

Students communicate ideas and describe models.

Students evaluate choices and engage in debate.

Students construct new hypotheses.

Students integrate new ideas with existing knowledge.

Students monitor their own ideas and knowledge organization.

**APPLICATION**

Students apply knowledge to new situations.

Students apply knowledge to previous experiences.

Students ask new questions.

Students take action.

Note. Adapted from Jones, (1994). Used with permission.

---

National Science Teachers Association

most appropriate measure of the constructivist teacher's goals. Objective tests typically measure knowledge that *the teacher feels is important* and may not measure what students know or understand. This conflict between constructivist teaching and traditional assessment is one reason that the movement toward alternative assessments is growing so quickly.

Performance-based tests and open-ended assessments can provide students with rich problems that allow them to utilize a variety of approaches and skills as they seek solutions. Student interviews have become an ideal way for teachers to gain understanding of their students' knowledge. Portfolios, another alternative assessment, allow students to become responsible for their own growth and to decide how to represent their achievements and understandings. Each portfolio will likely be different, just as each student's conceptual growth is unique.

One of the most interesting alternative methods of assessing student conceptual growth involves the use of concept mapping (Novak & Gowin, 1984). This technique involves having students brainstorm concepts on a selected topic and then arrange these concepts hierarchically such that the more inclusive concepts appear near the top of the concept map and the less inclusive concepts are placed below. Relationships between these concepts are then considered, and linking lines are drawn and labeled to indicate how one concept is related to another. Concept maps are not only a useful tool to measure how knowledge changes—they can also be a valuable metacognitive tool to help students think about their thinking.

Alternative assessments such as concept maps, portfolios, and interviews are not particularly helpful to educators who want to compare one student to another. Instead, alternative assessments are useful in assessing individual conceptual understandings.

## THE CONSTRUCTIVIST STUDENT

Just as the role of the teacher changes under constructivist philosophy, so does the role of the student. The goal of the constructivist classroom is to promote cognitive growth for each individual student. As a result, it becomes the student's responsibility to be an active participant in the meaning-making process. According to Ausubel (1968), meaningful learning requires a meaningful learning task that is relevant to the learner's cognitive structure; and, perhaps most importantly, there must be deliberate intent on the learner's part to relate the material in a nonarbitrary way to what is already known. This places the responsibility for learning directly on the student. Teachers cannot force students to learn, nor can they pour knowledge into a student's head. It is only through active participation in the learning environment that students can construct new knowledge that is relevant and meaningful. This process involves student questioning of readings, experiences, and peers, as well as of the teacher. Students must learn to clarify ideas and negotiate meanings. Metacognitive skills are essential for students to monitor their cognitive growth. The roles and responsibilities, as well as the teaching-learning process, are dramatically different for teachers and students in a constructivist classroom.

## THE CONSTRUCTIVIST SUPERVISOR/LEADER

How can science leaders provide leadership, promote change, and improve science programs? The roles and responsibilities for leaders in science education are numerous and perhaps overwhelming at times. Responsibilities may include establishing budgets, managing science equipment, designing evaluation procedures, developing curriculum, ordering and selecting textbooks, mentoring teachers, setting program goals, and initiating new innovations. Of the many tasks of the supervisor, team leader, or department head, the most difficult for many leaders are those that involve promoting the professional growth of teachers. Constructivist leaders cannot *make* change happen with people. Instead, constructivist leaders, like constructivist teachers, work collaboratively with others to negotiate and build common understandings. This process is difficult because of the variety of experiences, ideas, and content expertise that individual teachers hold. Regardless, leaders must value the diversity of ideas, values, skills, and experiences that teachers bring to schools. Through exploration of differences, leaders and teachers can negotiate common visions about goals for science programs.

At the heart of teacher development is the growth of pedagogical and content knowledge and skills. In order for teachers to experience conceptual change, Posner, Strike, Hewson, and Gertzog (1982) argue that four conditions must be present:

1. There must be dissatisfaction with existing conceptions.

2. A new conceptual idea or scheme must be intelligible.

3. New concepts must be plausible.

4. A new concept should be fruitful.

## STRATEGIES TO PROMOTE CONCEPTUAL CHANGE

Science education leaders can play a critical role in staff development by providing the perturbation, or cognitive conflict, that evokes dissatisfaction with existing concepts (Posner, Strike, Hewson, and Gertzog's first condition for conceptual change). This mismatch between proposed and existing views can be promoted through a variety of individual and collaborative experiences. Not all teachers can be expected to experience growth in the same way or at the same rate. Recognizing that there are multiple ways of knowing how to teach and learn science re-

sults in a more flexible and personal style of leadership.

For several years the School of Education at the University of North Carolina has been involved in an intensive program of constructivist staff development for elementary and middle school science teachers. This program, known as Constructivist Science, is based on constructivist theory and is designed to improve teachers' physical science content knowledge, as well as to promote the use of constructivist teaching in area schools. A number of strategies have been utilized that can be valuable in promoting teachers' cognitive change. These include student interviews, concept mapping, journal reflections, research readings, peer coaching, extensive peer-peer discourse, and portfolio development. Several of these strategies are described in the sections that follow; examples are provided that illustrate how each can promote professional development.

## STUDENT INTERVIEWS

Analysis of the reflections of teachers involved in our project have shown that conducting student interviews can be an extremely valuable method of promoting teachers' cognitive growth, relative to both pedagogy and science content knowledge. Many elementary and middle school teachers are genuinely afraid of revealing how little they know about science phenomena. Beginning with a focus on what students know has proven to be a non-threatening vehicle that leads to discussion and examination of teacher knowledge. Teachers are generally fascinated with their students' conceptual understandings, as revealed by the journal entry written by one participant in the project:

I am really enjoying the interview process. My kids also get very excited when I take them out to the library and tape our interviews. They love the attention and the opportunity to share their thoughts ... It is exciting to examine and question my beliefs and those of my students. I feel we are learning together. (Mary) [Pseudonym]

Another teacher described her surprise at seeing how her students described their understandings of science concepts after she had taught the unit:

It has been "eye opening" as I have assessed the students. Various misconceptions have surfaced that I would not have looked for or been aware of previously. I have been "rudely awakened" to find some of the concepts were not understood; and I thought we had done a great job. So, just when I thought I had things under control, I find there is another way—maybe better! (Susan)

## JOURNALS

Journal writing has for many years been viewed as an effective way to promote teacher reflection and has proven to be a useful metacognitive tool. For the constructivist leader, sharing journals can be a valuable way to gain insight into individual perspectives and contexts. For the staff development facilitator, journals provide information on the implementation of innovations that might not be gained through any other method. Judy, one of the teachers in the project, offered the following comments in her journal:

I have realized that my teaching methods are changing drastically. I used to give pages of notes on the overhead for the children to copy and memorize. For some reason I thought that taking notes would help the children understand the concept being taught. Now, however, I realize I was probably

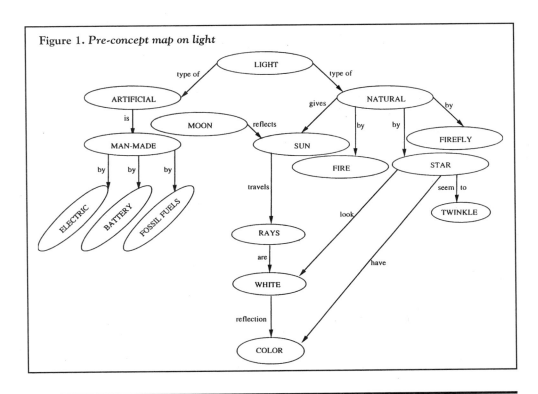

Figure 1. *Pre-concept map on light*

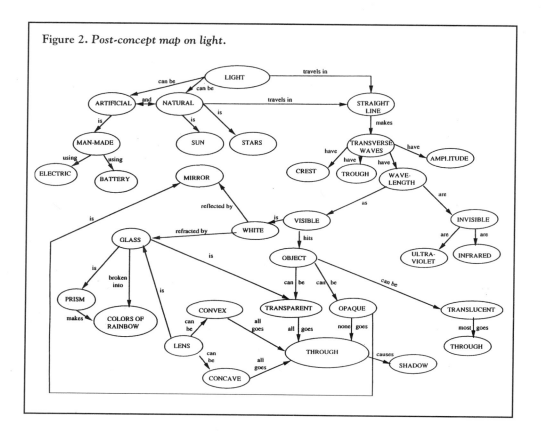

**Figure 2.** *Post-concept map on light.*

wasting a lot of the children's time. Even though we were involved in more projects at the time, there wasn't as much questioning being conducted.

Without sharing journals, the instructors would have been unaware of Judy's shift in her approach to teaching science, because she had never made these comments verbally during the project.

Another teacher, Heidi, shared the following reflections in her journal:

Something interesting from my classroom—I had my kids create concept maps on space and they were great. *It is easy to forget that the job of a teacher isn't always to shovel knowledge out to the kids—but to see what they know.*

*The thing that stands out most to me is that I am now practically "entranced" with children's conceptions/misconceptions and have really tuned in to my students much more now* [italics added]. I knew that children came to me with knowledge, of course ... I never truly stopped and took the time to listen to the "theories" they possessed. Looking back at that, I want to kick myself.

## CONCEPT MAPS

Concept mapping has emerged from the movement toward teaching for understanding as an interesting cognitive tool. In the Conceptual Science program, concept mapping was used as a repeated strategy for teachers to look at their conceptual change before and after studies of different science content areas (See Figures 1 and 2). Teachers reported that the concept mapping really made their knowledge, and lack of knowledge, about science explicit. Sharon, an elementary teacher, described the value of concept mapping in her journal:

Of the topics that we have explored, sound seems to be the hardest for me to get a real understanding of. Oddly enough, it is the topic that I thought that I knew the most about, but when I started to do a concept map, I didn't have much to write down. After a couple of classes on sound I have some of the vocabulary down, but I still feel that I have a very elementary knowledge of sound. When trying to redo my concept map on sound, I found that I brainstormed a relatively long list of words, but as I tried to place them and [figure out] the connecting phrase or word to use I realized the real benefit of the concept map. It enables you to map out exactly what you know and in the course of that mapping, you can see (because it is right in front of you) what you know. After I had this revelation I started to think about how I might make the concept map a useful tool in my classroom.

Concept mapping, journal writing, reading research, and interviewing students have been shown to be effective in promoting cognitive growth. Some strategies, such as concept mapping and interviewing students, are useful in encouraging teachers to experience conflict or dissatisfaction with existing conceptions. Others, such as reading, research, and peer coaching are useful in helping teachers see how new concepts can be plausible and fruitful. Discourse with colleagues is also helpful for teachers to see how new ideas can be applied to real-world classroom contexts.

## CULTURE AND CLIMATE ISSUES

It is critical to note that the process of conceptual change takes time. It is also more likely to occur if leaders create an atmosphere of trust and acceptance, as well as offer a willingness to admit that there are no right answers, only viable ones. The image of a lock and key is a useful metaphor: "Knowledge … fits our experience as a key fits a lock. The key enables us to open the lock but tells us little, if anything, about the lock. We all know that different keys can open the same lock" (Bettencourt, 1993, p. 43).

Creating trust between teachers and science education leaders is essential if meaningful change is to occur. The traditional power relationships based on coercion and domination are philosophically incompatible with constructivist philosophy. If we truly believe that each person constructs his or her own knowledge, if we value diversity, and if we acknowledge that no one person holds the keys to truth, then the sources of power for leaders must be derived not from external authority but, rather, from our abilities to create shared experiences where we can negotiate meaning and set common goals.

## LEADERSHIP FOR THE TWENTY-FIRST CENTURY

As we move into the 21st century, it is difficult to imagine, much less predict, what leadership skills will be needed. There have never been more science curricula, materials, resources, and safety issues with which leaders must contend. School organizational configurations continue to expand as school systems try new features such as block scheduling, interdisciplinary instruction, magnet schools, and year-round schools. The variables and problems that emerge from differing purposes and designs make predicting leadership needs nearly impossible. However, past studies have revealed that effective leaders hold common characteristics of personal development, empowerment of others, vision, communication skills, and trust (Bennis & Nanus, 1985).

Future leaders will find that focusing on self-development and professional growth will be crucial in order to keep up with the exponential growth in knowledge, technology, and educational reform movements. Studies of leadership have shown that effective leaders know themselves, their strengths, and their weaknesses (Rogus, 1988). They work to increase their understandings of science, education, colleagues, and students. The constant renewal of skills and knowledge supports the other characteristics of leadership—empowerment, vision, communication skills, and trust.

Empowerment of others is the key characteristic of constructivist leadership.

> Effective leaders work in transformational ways. They address themselves to followers' wants, ends, and other motivations, as well as their own, and thus they serve as a force in changing their followers' motive(s) ... Transformational leadership occurs in a way that leader and follower raise one another to higher levels of motivation. Their purposes become fused. (Rogus, 1988, p. 49)

Empowering others requires that leaders help others develop their fullest potential. Leaders who give away their power and authority to help others find their inner voice and direction, ultimately strengthen the entire system.

Vision, another characteristic of effective leaders, has been described by Rogus (1988) as "a mental journey from the known to the unknown, creating a future from a montage of hopes, dreams, facts, threats, and opportunities" (p. 49). The visions for science programs should emerge from the people, community, and culture of those who are involved. Visions are useful only if they are collectively created and shared. Assisting teachers, parents, and students in shaping visions of new and different science programs is a major challenge for the science leader.

Communication skills, accompanied by strong interpersonal skills, enable effective leaders to create a sense of community in which visions can be shared. In some cases, leaders must open the communication pathways and negotiate relationship building in order for colleagues to work together to set goals.

Finally, trust sits as a cornerstone supporting the relationships between leaders and those they serve. Trust is not easily developed, and is quickly lost if not nurtured. Developing trust takes considerable time and patience (Hickman & Silva, 1984).

As we look toward the 21st century, it is apparent that the characteristics of effective leaders must emerge from us all.

> Each individual must assume responsibility for improving science education through educating science teachers, developing curriculum materials, supervising state and local reforms, researching teaching practices, and most importantly, teaching science to boys and girls, early adolescents, and young adults who will be scientists and citizens. (Bybee, 1993, p. 9)

## References

Ausubel, D. (1968). *Educational psychology: A cognitive view.* New York: Holt, Rinehardt & Winston.

Bennis, W., & Nanus, B. (1985). *Leaders: The strategies for taking charge.* NY: Harper and Row.

Bettencourt, A. (1993). The construction of knowledge: A radical constructivist view. In *The practice of constructivism in science education,* by K. Tobin (Ed.), pp.39–50. Washington, DC: American Association for the Advancement of Science.

Brooks, J. (1990, February). Teachers and students: Constructivists forging connections. *Educational Leadership*, 68–71.

Bybee, R. (1993). Leadership, responsibility, and reform in science education. *Science Educator, 2*, 1–9.

Driver, R., Asoko, H., Leach, J., Mortimer, E., & Scott, P. (1994). Constructing scientific knowledge in the classroom. *Educational Researcher, 23*, 5–12.

Hickman, C., & Silva, M. (1984). *Creating excellence: Managing corporate culture, strategy and change in the new age*. NY: New American Library.

Jones, M.G. (1994). Constructing knowledge of science concepts. *NCSTA Journal, 3*, 13–16.

Lawson, A., Abraham, M., & Renner, J. (1989). *A theory of instruction: Using the learning cycle to teach science concepts and thinking skills*. (NARST Monograph No. 1).

Nadeau, R., & Desautels, J. (1984). *Epistemology and the teaching of science*. Ottawa: Science Council of Canada.

Novak, J., & Gowin, D. (1984). *Learning how to learn*. Cambridge: Cambridge University Press.

Posner, G., Strike, K., Hewson, P., & Gertzog, W. (1982). Accommodation of a scientific conception: Toward a theory of conceptual change. *Science Education, 66*, 211–227.

Rogus, J. (1988). Teacher leader programming: Theoretical underpinnings. *Journal of Teacher Education, 39*, 46–52.

von Glasersfeld, E. (1993). Questions and answers about radical constructivism. In *The practice of constructivism in science education*, by K. Tobin (Ed.), pp.23–38. Washington, DC: American Association for the Advancement of Science.

## Author Note

**M. Gail Jones** is an associate professor at the University of North Carolina at Chapel Hill. Her research focuses on learning in a sociocultural context. She has authored books and articles on middls school science, assessment, constructivism, gender equity, and estuarine ecology. Her research has received awards from the Association of Educators of Teachers of Science (AETS) and the Association for Supervision and Curriculum Development (ASCD).

# Moving From Administration To Leadership: Science Leaders In A Changing World

John Wallace
Helen Wildy

The world has changed dramatically since the publication of the first version of this *Sourcebook* in 1967. In many places, schools and school leaders are struggling to respond to the changes being asked of them. Notions of how schools and classrooms are structured, organized, and led face increasing scrutiny. Similar criticisms are made of traditional approaches to subject matter and pedagogy. Schools and their leaders are asked to become more highly responsive and adaptable to changing circumstances and requirements. School communities, it seems, need to reconceptualize the means by which they go about the task of educating people. By and large, schools and classrooms have retained the same basic structures for the past 50 years and frequently are too rigid to respond to the exponential growth of knowledge in the world around them. Nowhere is the need for change more evident than in school science. Alvin Toffler's (1990) phrase "a constipated approach to knowledge" colorfully describes how many organizational arrangements inhibit the growth of scientific knowledge, skills, and understanding in young people. And if schools are to move beyond these inhibiting ways of working, then new visions based on new beliefs are needed. The administrators and managers of traditionally organized structures must give way to leaders of innovative dynamic science programs.

## CHANGES IN SCHOOLS

Many current practices in schools and classrooms come from entrenched beliefs about science and about the world in which educators work. In seeking more relevant ways of thinking about schools, we need to challenge beliefs about the nature of science, the nature of teaching, and ways of working together.

First, views about the *nature of science* are changing. Once, it was generally accepted that science consisted of a set of universal truths that described the operation of the natural world. Now, many believe that science is a process of personal sense-making that helps us survive in our environment. And associated with these contrasting views about science are beliefs about the purposes of school science. Where once schools were required to deliver a set of uniform goals for school science, now those responsible for science must allow the possibility of multiple, and sometimes competing, goals.

Second, views about the *nature of teaching* are changing. Once, it was thought there was one best way to teach science. The science curriculum was packaged into digestible pieces of content, ready to be dispensed by the teacher when appropriate. Now, there is an understanding that teaching science requires a wide range of strategies tailored to suit the needs and prior knowledge of individual students.

Third, views about the most effective *ways of working together* are changing. Once, it was accepted that power in schools and school systems should be located at the top. Now, it is agreed that power in schools is better shared among all members of the school community. In the past, the school community included a public that was generally supportive of school science in marginal but predictable ways. Today, not only does the public influence school science in major ways, but those responsible for school science have to deal with the diversity and unpredictability of that influence. Where once the science administrator made decisions alone, now those responsible for science in schools must make decisions collaboratively to satisfy the competing agendas of all those involved.

## FROM ADMINISTRATION TO LEADERSHIP

Changes in schools such as those described above have important implications for the work of the science leader. In the past schools may have been well served by a school administrator, but the changed educational environment now calls for strong educational leadership. And for leaders to be effective in this new environment, they need to:

• develop *trust* among their colleagues by allowing them choice rather than controlling them through compliance;

• focus on *improvement*, not just maintenance;

• work in *collaboration* rather than in isolation;

• foster *responsibility* rather than dependency among teachers;

• *facilitate* problem solving rather than solve problems for their colleagues; and

• *listen* more than tell their colleagues what to do.

## Trusting (Rather than Controlling)

Educational leaders develop relationships among teachers that are characterized by trust rather than by control. They allow their colleagues choices about how to do their work rather than demanding compliance to rules and regulations. When science teachers are trusted to make their own decisions rather than relying on imposed rules and procedures, those decisions are more finely tuned to local circumstances.

But there is an even more powerful reason for trusting teachers. Teachers live in a rapidly changing environment, with a constantly expanding knowledge base, increasing complexity of practice, and pressure from legitimately conflicting goals. In such an environment, teachers are called upon continually to make decisions about appropriate courses of action. These decisions cannot be made to fit within rigid rules of conduct. Because teaching as a profession recognizes the individuality and uniqueness of those who teach and those who learn, compliance with uniform and standardized processes and procedures is inappropriate. Neither more precise prescription, nor total discretion, is the answer. What is needed is an emphasis on continual learning that helps teachers reflect on the outcomes of their actions for others. And so a key function of the science leader is to foster reflective learning opportunities for teachers within a climate of trust.

## Improving (Rather than Maintaining)

Too often, schools and school systems are dogged by a bureaucratic mind-set. The bureaucracy itself is seen as an obstacle to change. We often find people saying, "You can't do that because of the schedule" or "That's not within the guidelines." Focusing on their own classrooms and employing a narrow information base, teachers often feel incapable of changing their situations. Even collaborative improvement efforts are thwarted because the structures around them are turned into inflexible systems of bureaucratic maintenance and control.

On the other hand, schools that focus on improvement have a broad information base. Individuals collaborate with their colleagues. They strive for better results and learn to become more effective. Science departments operate as professional communities; leaders and teachers are seen as career-long learners. Learning and performance are intimately related; the high performers are those who learn most quickly.

What we *say we do* needs to be consistent with *what we do*. When schools are primarily concerned with bureaucratic issues, teachers focus on repetitiveness, efficiency, predictability, and quantity. A top-down approach, a short-term focus, fixed job duties, and low member participation are characteristics of such schools. By contrast, in professional science departments, the goals are quality and innovation. The behaviors promoted reflect personal responsibility, problem solving, creativity, flexibility, and constant improvement. Science leaders develop ways of working together that feature a participative approach, a long-term focus, development, and support.

## Collaborating (Rather than Working in Isolation)

How people work and relate to each other differs in bureaucratic and professional organizations. Rosenholtz (1989) distinguishes between "stuck" and "moving" schools. She found that stuck schools were bureaucratic in their structure. They were not support-

ive of change and improvement; uncertainty and isolation went hand in hand. One of the main causes of uncertainty, Rosenholtz found, was the absence of positive feedback. In their professional isolation, teachers and their leaders neglected each other; they overlooked the importance of complimenting, supporting, and acknowledging each other's positive efforts.

By contrast, in moving schools, teachers worked in a professional manner. Even the most experienced believed their job was inherently difficult. Almost everyone recognized that they sometimes needed help and could learn from others in the department— giving and receiving help was accepted practice. Having their colleagues show support and communicate more with them about what they did helped teachers grow in confidence. As a result, they became more certain about what they were trying to achieve and how well they were achieving it.

Schools that work collaboratively respect, celebrate, and recognize the teacher as a person and as a professional. Science teachers are empowered to define what counts as a problem. They are also empowered to work together to change the conditions that caused the problem. Developing collaborative cultures is a complex but critical function of science leadership.

## Developing Responsibility (Rather than Dependency)

There is a quiet revolution in innovative organizations away from tighter controls, precisely defined jobs, close supervision, and dependency, and towards a more entrepreneurial spirit. This spirit is typified by increased autonomy, responsibility, and professionalism. It has been found that people work best as members of a team where leadership, knowledge, and responsibility are shared and where individuals function as equals without threat to one another.

Encouraging science teachers to take responsibility for the way they do their work has significant advantages for schools as learning communities. First, it empowers teachers to solve problems and undertake initiatives related to their work environment. Second, it diffuses power and responsibility for decisions throughout the school. Third, it reduces teachers' dependency on their leaders, enabling leaders to shift their energies from routine bureaucratic tasks to educational issues.

Accepting responsibility for professional performance also means a new style of accountability in relationships between science leaders and teachers. Decisions about how teachers discharge their responsibility rest with the teachers themselves, not with the person to whom they are accountable. How people develop and engage in their accountability relationship depends on the way they talk to and treat each other. Perhaps the most challenging part of the process is the way feedback is sought and given. To improve performance, both people in the relationship must be open to each other's messages. And it seems that feedback is more likely to have an impact when it is ongoing, related to what is observed, solicited rather than imposed, tied to behavior rather than to personality, and occurring on a day-to-day basis rather than once or twice over a long period of time.

## Facilitating Problem Solving (Rather than Solving the Problems)

Administrators often control the flow of information and direct the work of others. Invariably, they act as gatekeepers, solving—or failing to solve—problems on behalf of others in the organization and fostering a sense of dependency on the leader. In professional organizations, however, the power is shared across the community. Schools that operate this way are more flexible and capable of responding appropriately to problems as they arise. Leaders support problem-

solving and initiative-taking by others. They set up cross-hierarchical groups of teachers, leaders, and sometimes students and parents. They delegate authority and resources to the groups while maintaining active involvement in, or liaison with, the groups.

Teachers become empowered when they can solve their own problems. In science departments that are professional communities, teachers have access to information and resources enabling them to make decisions. Most significantly, they can count on the support of their science leader. In such communities, leaders facilitate a constructive working environment where new ideas and experimentation are valued.

## Listening (More than Telling)

There is little doubt that professional communities place a large emphasis on enhanced communication. Effective leaders use a variety of communication styles and responses. One continuum of possible reactions by a science leader to a teacher includes:

- not listening

- listening

- listening and hearing

- listening, hearing, and encouraging the teacher

- listening, hearing, encouraging, and agreeing to share the risks and consequences of the proposed initiative. (Barth, 1989, p. 240)

This continuum reflects an increasing level of trust among members of the department. It reflects the belief that all staff have a shared responsibility in the delivery of the science program. Capable communicators use an appropriate range of communication styles. They tend to involve people by sharing their excitement, aspirations, and goals. They put their energy into trying to understand other people's perspectives, rather than telling them what to do.

## HELPING OTHERS TO GAIN SKILL

We argue that the world has changed, and so must schools. However, changing schools and school science means changing people, not an activity to be undertaken lightly. Many school leaders, often experienced in old ways of working, find shifting to new ways creates difficulties. It is not surprising that science leaders, like all school leaders, continually face resistance in themselves and in others when trying to adopt new skills and alter familiar tasks. Part of this process involves changing from telling people what to do or doing it themselves, to helping others gain skill.

Remember that science teachers are also leaders. People in positions of authority are members of teams, jointly making decisions and sharing authority. It is everyone's responsibility to share leadership. This benefits the group as a whole because it develops the expertise, skill, confidence, and commitment of all members of the team. Most importantly, science leaders need to demonstrate integrity. A leader who demonstrates integrity inspires trust in others. The leader will not have to constantly prove the good intent of his or her actions. Further, if integrity is a guiding principle, then the leader can trust others. This, in turn, opens the door to shared decision making, teamwork, and empowerment—basics for a professional community of science teachers.

To set up new ways of working together we need to be patient and forgiving, just as we have been tolerant of the shortcomings of old ways of working. We cannot seriously expect a utopian substitute for a bureaucratic administration. Nevertheless, it is possible to develop a viable and realistic alternative that trades control for freedom and that is responsible and responsive to the learning needs of all of our students.

We began this chapter with reference to the first version of this publication which included the term "science supervisors" published in 1967. Our preference is for the term "science leader". The difference between these two terms is more than superficial. This reflects a fundamental shift in our understanding of leadership, schools, and science teaching, and of the kind of skills required to lead our science programs into the next century.

## References

Barth, R.S. (1989). The principal and the profession of teaching. In *Schooling for tomorrow—directing reforms to issues that count*, by T.J. Sergiovanni & J.H. Moore (Eds.). Boston: Allyn & Bacon.

Rosenholtz, S.J. (1989). *Teachers' workplace: The social organization of schools*. New York: Longman.

Toffler, A. (1990). *Powershift: Knowledge, wealth and violence at the edge of the 21st century*. New York: Bantam.

## Author Note

**John Wallace** is senior lecturer at the National Key Centre for School Science and Mathematics at Curtin University of Technology, Perth, Australia. His career includes 20 years of experience as a science teacher, school administrator and science supervisor. He has published over 100 articles, chapters, and books on science education and educational leadership.

**Helen Wildy** is a research associate at the National Key Centre for School Science and Mathematics at Curtin University of Technology, Perth, Australia. She also works as an educational consultant in the field of school leadership and has published extensively in the area.

# Bridging The Gap Between The Old And The New: Helping Teachers Move Towards A New Vision Of Science Education

Charles R. Barman

"Certainly, the greatest enemy of understanding is coverage— the compulsion to touch on everything in the textbook or the syllabus just because it is there ... (p.203)" In this quote, Howard Gardner and Veronica Boix-Mansilla (1994) describe one the most pervasive problems of our current system of education. Project 2061 (Rutherford & Ahlgren, 1990) and *National Science Education Standards* (National Research Council, 1996) address this same issue. Both curriculum reform efforts recommend teaching for science understanding. To accomplish this goal, members of both efforts believe students will need to study fewer topics in greater depth. In addition, these projects advocate using teaching models that engage students in active learning. Students are encouraged to describe objects and events, ask questions, construct explanations of natural phenomena, test those explanations, and communicate their ideas to others.

As schools move toward developing curricula that are consistent with the goals of Project 2061 and *National Science Education Standards*, many teachers may need to modify their science materials and their delivery of these materials. There are techniques that can help teachers who fall into this category meet the current trends in science education.

This article will outline a procedure that has helped both preservice and inservice teachers modify their science teaching to be consistent with current science education goals (Barman, 1992; Barman & Shedd, 1992; Barman & Shedd, 1993). However, before describing this procedure, there will be a brief discussion of the teaching model that was introduced as part of this pedagogical shift.

## THE LEARNING CYCLE

The teachers used a teaching model called the learning cycle. This instructional model has been shown to be effective at a variety of grade levels and with a variety of educational materials (Purser & Renner, 1983; Saunders & Shepardson, 1987; Stepans, Dyche, & Beiswenger, 1988; Barman, 1990; Marek & Methven, 1991). The learning cycle originated in an elementary science program known as the Science Curriculum Improvement Study (SCIS) and consists of three distinct phases: (a) exploration, (b) concept introduction, and (c) concept application (figure 1). During the *exploration phase*, the teacher presents a task or problem that is open-ended enough to encourage students to

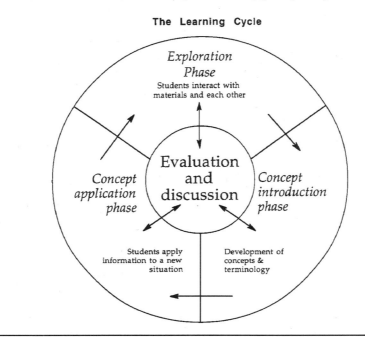

Figure 1. *Unidirectional arrows indicate the relationship between the phases of the learning cycle or how one phase leads to the next one. Ideally, the concept application phase of one lesson can lead to the exploration phase of a new lesson. The bi-directional arrows indicate that evaluation and discussion can be integrated into any part of the cycle.*

**The Learning Cycle**

*Exploration Phase*
Students interact with materials and each other

Evaluation and discussion

*Concept application phase*

*Concept introduction phase*

Students apply information to a new situation

Development of concepts & terminology

pursue a range of strategies, yet specific enough to provide some direction. This first phase engages students in a highly motivating activity as the basis for developing a specific concept and related vocabulary. It also allows the teacher to explore with students their existing ideas about the concept and to identify any inaccuracies in their views of this concept or related concepts.

In the second phase, *concept introduction*, the teacher gathers information about the exploration experience from the students and uses this to introduce the main concept of the lesson and any new vocabulary related to the concept. Materials such as textbooks, audio-visual aids, or other written materials may be used to facilitate the concept introduction.

The final phase, *concept application*, is an opportunity for students to study additional examples of the main concept of the lesson or be challenged with a new task that can be solved on the basis of the previous exploration activity and concept introduction. Ideally, the additional examples of the concept or the new task will have a direct relationship to the everyday lives of the students.

## HELPING TEACHERS USE THE LEARNING CYCLE

According to Bandura (1986), the most effective way to have individuals feel confident about implementing a new technique or idea is to first have someone model the technique and show how it can be put into practice. After sufficient modeling has occurred, the individuals are asked to apply it to their own situations.

Basically, the procedure this section describes follows the steps identified by Bandura. For example, the learning cycle is introduced to teachers via good modeling. They are provided with several science lessons that are presented in the learning cycle format. They are also asked to compare and contrast the learning cycle lessons with hypothetical lessons that present the same information in a different format. In addition, they discuss which approach they feel would be most effective in helping students understand the concepts presented in the lessons. Table 1 displays an example of a learning cycle lesson that is presented to teachers and two hypothetical lessons that deliver the same information but in a different format.

After the teachers have experienced several learning cycle lessons and have compared and contrasted them with other approaches, they are asked to analyze a portion of their science materials to determine whether its lessons follow the learning cycle

---

**Table 1.** *A Comparison of a Learning Cycle Lesson and Two Other Alternative Lessons*

Lesson Topic: Electric Circuits

*Learning Cycle Lesson* :

*Exploration Phase*

The students are asked to use a D-cell, 2 pieces of insulated wire, and 1 flashlight bulb to light the bulb. (*Caution*: The wire can get hot.) Then they are asked to make a drawing of a "bulb system" that lights the bulb and one that does not.

*Concept Introduction Phase*

Several students are asked to draw one of their bulb systems on the board, and the class is asked to predict whether each system will light. The students then read the pages in the textbook that introduce the terms *electric circuit, open circuit,* and *closed circuit*. Then each student is asked to draw an example of a closed circuit and an open circuit and explain how a bulb system works.

*Concept Application Phase*

The students are given a D-cell, a battery holder, 3 pieces of insulated wire, 2 tacks, a piece of cardboard, a strip of aluminum foil, a flashlight bulb, and a bulb holder. With these materials, they are asked to form a circuit with a switch that can turn the bulb on and off and describe how the switch opens and closes the circuit. Then they are asked to explain how the light switches in the classroom work. Each student is asked to make a drawing to illustrate how these switches open and close the light circuit.

*Alternative Lesson 1* :

The students are asked to read the pages in their science textbook that pertain to electric circuits, open circuits, closed circuits. After they have read this material, the teacher discusses it with them. The teacher also supplements the reading with a short lecture about how electrical systems work in most homes. Then the teacher asks the students to perform an activity in which they construct a open and closed circuit. The materials they use are a D-cell, a battery holder, insulated wire, a flashlight bulb, a bulb holder, and a switch.

*Alternative Lesson 2* :

The students are told that they are going to learn about open and closed circuits. The teacher shows them several posters that depict different examples of open and closed circuits. Then the students are asked to read about this topic in their textbooks. When they complete the reading, the teacher discusses it with them and assigns three written questions from their textbook for homework.

---

format. During this analysis, they are also asked to list any modifications that would be necessary to reconstruct the lessons to follow the learning cycle. The extent to which modifications are needed varies with each set of materials; however, in most instances, the teachers find that the lessons lack either good exploration or concept application activities. For example, Table 2 outlines a typical lesson found in most elementary science materials and identifies how it could be modified to conform with the learning cycle.

After discussing their analyses, the teachers are asked to select the part of their science materials that they intend to use next with their students and organize its lessons to follow the learning cycle. The teachers are encouraged to use the following steps to accomplish the task:

1. Examine the lessons to determine the extent to which they follow the learning cycle. Use the Learning Cycle Checklist (Table 3) to assist in this effort.

2. Supplement the parts of the learning cycle that are missing. The teachers are encouraged to use a variety of resources to help generate ideas for this task.

3. In peer groups, evaluate the lessons using the Learning Cycle Checklist. Workshop facilitators are also available to assist in this evaluation.

---

**Table 2. *Modifying Materials to Follow the Learning Cycle***

Lesson Topic: Complete and Incomplete Metamorphosis

*Current Presentation* :

The teaching materials contain information about the stages of complete metamorphosis. There are excellent illustrations to explain each of these stages. At the end of the presentation of both forms of metamorphosis, the text has an activity that has the students use hand lenses to observe different stages of a mealworm's life cycle (e.g. larva, pupae, adult).

***Learning Cycle Modification*** :

*Exploration Phase:*

Using hand lenses, the students observe several mealworms in different stages of their life cycle (e.g. larva, pupae, adult). Ask the students to list the physical characteristics of each specimen, and have them draw a picture of each one.

*Concept Introduction Phase*:

Discuss the students' observations and record them on the board. Introduce the terms *egg, larva, pupa, adult,* and *complete metamorphosis*. Explain that the specimens they examined were examples of the larva, pupa, and adult stages of a specific type of beetle. Ask the students to read the part of their textbook that discusses complete metamorphosis. Introduce the term *incomplete metamorphosis* and ask the students to read the part of their textbook that explains this phenomenon. Ask them to differentiate between complete and incomplete metamorphosis.

*Concept Application Phase:*

Display a variety of insect specimens (adult and immature organisms displaying characteristics of incomplete and complete metamorphosis), and provide the students with hand lenses. Ask the students to observe each specimen and predict what stage of development each one represents (e.g. larva, pupa, nymph, or adult). The students could verify their predictions with an insect identification guide such as [H.S. Zim's *Insects* (1951, Western Publishing Company, NY)].

---

National Science Teachers Association

**Table 3.** *Learning Cycle Checklist*

| *Exploration Phase* | Yes | No |
|---|---|---|
| A. The lesson contains an exploration phase that is activity-based. | ___ | ___ |
| B. Ample time is provided for the exploration phase. | ___ | ___ |
| C. The exploration activity provides student-student and student-teacher interaction. | ___ | ___ |

| *Concept Introduction Phase* | | |
|---|---|---|
| A. The concept(s) is named or appropriate vocabulary is developed after an exploration activity. | ___ | ___ |
| B. The concept(s) and term(s) are an outgrowth of the exploration phase. | ___ | ___ |

| *Concept Application Phase* | | |
|---|---|---|
| A. The students extend the concept(s) to one or more new situations. | ___ | ___ |
| B. Appropriate activities are used to apply the concept(s). | ___ | ___ |

Once the teachers have effectively modified the lessons to follow the learning cycle, they are asked to present these lessons to their students. A follow-up session is scheduled to allow the teachers to discuss the implementation of these lessons. They are also encouraged to develop peer teams in their schools to assist one another with the further development and critique of additional learning cycle lessons and to discuss the resolution of problems that may occur during the use of these lessons.

## CONCLUSION

As indicated above, this procedure has been used in a variety of teacher workshops and has proven to be an effective way to introduce teachers to the learning cycle and to assist them in applying this teaching strategy to their current science materials (Barman, 1992; Barman & Shedd, 1992; Barman & Shedd, 1993). It appears that the success of this procedure could be attributed to two factors. First, it follows the steps outlined by Bandura (1986) as those necessary for helping individuals become proficient in using a new technique. Second, it uses the same sequence that is found in the learning cycle. For example, the teachers' exploration activity consists of lessons that follow the learning cycle and some alternative lessons that use a different approach to present the same content. Their concept introduction includes comparing and contrasting the different lessons and discussing the benefits of each in developing concepts. Their application consists of having them analyze their own materials in relation to the learning cycle and modify these materials to follow this approach. In addition, they are asked to teach the modified lessons to their students and discuss the implementation of these lessons. In other words, not only are the teachers introduced to the learning cycle through sample lessons, but they are also exposed to this model through the overall design of the procedure itself.

In the last several years, teachers who have been exposed to this procedure and have used the learning cycle to present their science lessons have expressed very positive feelings about this shift in pedagogy. Some teacher quotes regarding this shift are as follows (Barman, Barman, & Cullison, 1991):

• Students modify their own concepts as they discover and learn via the learning cycle.

- Using the learning cycle, the students learned that they can get information from a variety of sources—not only teachers.

- Frequently, students learned more than expected from the planned lesson.

- The use of the science textbook within the context of the learning cycle changes from the traditional chapter-by-chapter approach to a more flexible approach.

- Rather than being textbook bound and teacher directed, science units become student centered and focus on the students' present knowledge.

The procedure described in this paper is one way to provide teachers with the assistance they need to modify their current teaching practices. Science leaders who elect to use this procedure will find that it is an inexpensive and effective way to help teachers develop more meaningful science lessons and, at the same time, move their teaching towards the new vision of science education.

## References

Bandura, A. (1986). *Social foundation of thought and action: a social cognitive theory.* Englewood Cliffs, NJ: Prentice Hall.

Barman, C., Barman, N., & Cullison, M.G. (1991). *Doing science using the learning cycle.* Indianapolis, IN: Indiana University School of Education.

Barman, C. (1990). *An expanded view of the learning cycle: new ideas about an effective teaching strategy.* (monograph #4). DC: Council for Elementary Science International.

Barman, C. (1992). An evaluation of the use of a technique designed to assist prospective elementary teachers use the learning cycle with science textbooks. *School Science and Mathematics 92,* 59–63.

Barman, C. & Shedd, J. (1992). An inservice program designed to introduce K-6 teachers to the learning cycle teaching approach. *Journal of Science Teacher Education 3,* 58–64.

Barman, C. & Shedd, J. (1993). A science inservice program designed for teachers of Hearing-Impaired children. In *Excellence in educating teachers of science,* by P. Rubba, L. Campbell, & D. Thomas (Eds.,)pp. 237–246. Columbus, OH: ERIC Clearinghouse for Science, Mathematics, and Environmental Education.

Gardner, H. & Boix-Mansilla, V. (1994). Teaching for understanding in the disciplines- and beyond. *Teachers College Record 96,* 199–218.

Marek, E. & Methven, S. (1991). Effects of the learning cycle upon student and classroom teacher performance. *Journal of Research in Science Teaching 28,* 41–53.

National Research Council (1996). *National Science Education Standards.* DC: National Academy Press.

Purser, R. & Renner, J. (1983). Results of two tenth-grade biology teaching procedures. *Science Education 67,* 85–98.

Rutherford, J. & Ahlgren, A. (1990). *Science for all Americans.* NY: Oxford University Press.

Saunders, W. & Shepardson, D. (1987). A comparison of concrete and formal science instruction upon science achievement and reasoning ability of sixth grade students. *Journal of Research in Science Teaching 2,* 39–51.

Stepans, J., Dyche, E. & Beiswenger, R. (1988). The effect of two instructional models in bringing about a conceptual change in the understanding of science concepts by prospective elementary teachers. *Science Education. 72,* 185–196.

## Author Note

**Charles R. Barman** is director of teacher education and professor of science and environmental education at Indiana University at Indianapolis. He has authored and co-authored several books and articles on science and science education. In addition, he has received a number of honors, including a NSTA Gustav-Ohaus Award and a Distinguished Teacher Educator Award for the State of Indiana.

# Mounting And Maintaining An Elementary Science Program: What Supervisors Can Learn From Research

Mary Budd Rowe

Research discussed in this paper will be helpful to supervisors who have some responsibility for science in the elementary school. It focuses on roles and activities of principals and teachers who are the direct beneficiaries of supervisor decisions and assistance. The paper begins with help and support needed by principals and then moves to that needed by teachers. Research on children's science learning appears only as it enlightens the principal and teacher agendas.

## THE PRINCIPAL COMES FIRST!

To cultivate and maintain a strong inquiry-oriented, activity-based science program, think first of the kind of information and support the principal needs. Rowe, McLeod, and Lawlor (1995) selected for study six elementary principals who had strong schoolwide science programs in place. Interviews with them and their teachers as well as video tapes of the principals as they "walked their talk" help us see the task of mounting and maintaining a strong elementary science program from the perspective of the principal. Although the schools' locations and clienteles differed considerably—rural, suburban, urban, east and west—the conditions that had to be met to maintain a successful program were the same across all sites. Administra-

tors pay particular attention to logistics, management, and program evaluation. They are interdependent factors. All three have to be well done all the time. If any one is weak, the program suffers.

## Logistics

Conventional wisdom holds that an elementary program usually fails because teachers do not know enough. Principals in the study say staff willingness to learn and experiment with a science program depends on two factors: the kind of support and understanding principals provide and how successfully the logistical demands of the program are met on a daily basis. Rowe et al (1995) point out that a school may have a good science curriculum and teachers who know the program content, but if principal and staff together do not establish an effective management system for distributing, collecting, maintaining, and replacing materials the program will fail. Supervisors can help principals recognize and plan to meet the logistical demands of an elementary program.

## Program Evaluation

Principals need help with how to do informal program evaluation. To do this they must go into classrooms. Many are office-bound captives of administrative demands not directly tied to curriculum or instruction. How much they learn from visiting classrooms, talking with students and teachers, and looking at use of materials will depend on their skills at interacting with children, their participation in the activities going on in the class, and their grasp of what the program is meant to accomplish.

All six principals in the study said they reserve an hour a day for visiting classes. Benefits they identified include at least the following: firsthand knowledge of what is going on; a better grasp of logistics; a collection of useful anecdotes for use in public presentations and with parents; more fruitful faculty meetings, because principals and staff both see the problems firsthand; early warning of problems in need of attention; and a student body and staff that recognize that the principal cares about what they are doing in science. For principals these are important aspects of program evaluation. It is informal, reflective, and inexpensive, and it contributes to improving program functioning.

Systematic, more formalized outcome evaluation probably should not begin until a program is functioning reasonably well. To begin outcome evaluation too early can distract attention from the main task of getting the program established. Formal evaluation can begin most fruitfully when the program is mature enough to meet three basic conditions suggested by Nicholson, Weiss, and Campbell (1994): a statement of desired outcomes exists, strategies for documenting these are set, and staff recognizes how program activities and teaching strategies relate to desired outcomes.

## Science Specialists

Administrators want to know whether they should hire a science specialist to teach science. There is no good research base yet on which to make recommendations. Four of the six elementary principals mentioned earlier had some person designated as a science re-

source teacher. All six principals warned that if a school has a science specialist the person needs to be thought of as a resource to be employed for helping faculty plan and teach science. They did not recommend that all the science teaching be left to one person. Instead planning and teaching should be collaborative. Otherwise the staff will not "own" the program, and the amount of science exposure will be limited to the number of classes the specialist can cover in a week. Moreover, connection of science to other content subjects will be less rich. In short, science periods in which a specialist participates must be treated as an enhancement opportunity that connects to the ongoing science program at each grade level.

Hope of developing and maintaining a successful elementary science program depends on creating conditions in which adults find it interesting and fruitful to teach science. Supervisors can help administrators and teachers create those conditions. Because so many supervisors come from secondary school backgrounds, it may be well to remember that unlike their high school colleagues, elementary teachers have numerous contents to teach. They do not have the relative luxury of dealing with only one or two fields of inquiry, and in many states they are besieged by new content standards in each of the disciplines: science, social studies, mathematics, language arts, geography, etc. They are getting very little help putting them all together.

## ACTIVITY-BASED, INQUIRY-ORIENTATION

There does seem to be a common theme that permeates all the standards, namely, students as inquirers and/or problem solvers. In a review of research on achievement in process-oriented science curricula in the elementary school, Barr (1994) found activity-based, process-oriented programs produced better problem solving achievement; students in such programs had better achievement than students in textbook-oriented programs; disadvantaged students appeared to gain the most from activity-based programs; and process-based curricula did not sacrifice content. She reported that regardless of teacher experience, students in activity-based classrooms outperformed students in more traditionally taught elementary classrooms.

The fact that teachers need materials for science instruction underlines the importance of logistics, mentioned earlier, to the success of an activity-oriented science program. Nothing is more discouraging to a harried, hurried teacher than to turn to a kit of materials and find it partially decimated, full of dirty glassware, or short of expendables needed for a given activity.

## PEDAGOGY: KEEPING CONTROL WITHOUT SUPPRESSING IDEAS

Attitudes, beliefs, and feelings teachers have about what science is and what it should accomplish will influence their ways of carrying out instruction. If the meaning they attach to inquiry and problem solving in science implies finding right answers, then they will focus on teaching and testing for facts. They will be less interested in making inferences and constructing explanations based on data, because it is not part of their vision or personal experience of science. If, however, they hold an alternative, more comprehensive vision of science, they will encourage discussions, dialog among students, and comparisons of competing explanations. In both views facts are important, but in the first they are an end product of instruction, while in the latter they form part of the data to be interpreted. People may agree on facts, but they often disagree on what the facts mean. In the less traditional view, experiments are a way of asking questions of nature, and science is a special form of making stories about how the world works.

Some pedagogical procedures that support a story-building approach to science are

well known and have a good basis in research. They include cooperative small group work doing investigations as well as much conversation within and between groups that makes members aware of differing perspectives and stimulates further inquiry as they seek clarification. See, for example, Johnson & Johnson (1987); Cohen (1984); Cohen & Lotan (1991).

Implications for inservice seem clear. Three interdependent elements must receive attention: existence of a curriculum that specifies the science content, design of activities meant to help students explore ideas, and pedagogical strategies suitable to the situation. If any one of these factors is poorly completed, the whole structure suffers. If, for example, the curriculum calls for an inquiry approach to a topic but the instructional design for laboratory activities is "cookbook" and the teaching (pedagogy) is drill and practice, then the quality of the program suffers. Teachers need to experience science initially on topics that are interesting for them as adults, and they need to do that in group work and discussion contexts.

Cohen and Intili (1982) found that the greatest achievement gains came in classrooms of teachers who were successful in arranging things so that children could talk and work together at multiple learning centers. The goal is to give students as much opportunity as possible to hold task-oriented conversations with each other about what they are doing and thinking. For teachers this amounts to delegating some authority to students to determine how to pursue their investigations.

Some teachers have trouble understanding the difference between delegated authority and laissez-faire. Too often in teaching science, teachers fear they will lose control. This is particularly likely when they do not understand the distinction between management of behavior and management of ideas. Loss of control during a science class can happen. When children are excited and a multitude of materials need to be distributed efficiently but the appropriate group skills are not in place to ensure a reasonably ordered work environment, the result can be chaos. Cohen and DeAvila (1983) developed an effective method for training teachers to delegate authority in cooperative group settings. Students learn how to perform certain roles necessary for effective collaboration on tasks that demand a substantial degree of interdependence, e.g., facilitator, equipment manager, etc. They internalize norms relating to behavior in inquiry-oriented cooperative settings. For instance, anyone has the right to ask for help; everyone is obligated to give help.

## Conversation *vs.* Inquisition

Classroom discussions that have more the character of conversation than inquisition produce gains in the length and quality of student reasoning. A long chain of research initiated by Rowe (1974) and continued by many others (see Rowe, 1986 for a review) points to the value of encouraging teachers to increase wait time from an average of one or two seconds to an average of three to five seconds at two points in classroom discourse: after they ask a question and after a student makes a response. If teachers also reduce the number of verbal rewards, then the combination of longer wait times and fewer verbal rewards produces seven outcomes supportive of an inquiry-oriented pedagogy: the length of student responses increases, students do more volunteering and participation rates by a greater number of students increase, failures to respond decrease, confidence increases, students ask more questions of each other and of the teacher, disciplinary moves decrease, and students do more speculating, i.e., consider a greater variety of possible explanations. All of these outcomes imply, of course, that questions on the part of the teacher are stimulating and the environment

of discussion invites alternative story lines to develop and be challenged.

As Hannah Arendt reminds us in *Men in Dark Times*,

> However much we are affected by the things of the world, however deeply they may stir and stimulate us, they become human for us only when we can discuss them with our fellows … We humanize what is going on in the world and in ourselves only by speaking of it, and in the course of speaking of it we learn to be human. (p.24)

Clearly, children and teachers need time and circumstances in which to talk together. We need to be aware, however, that opportunity to do and talk science with the teacher or with fellow students often is not equally available to all students in a class. In two reviews of research related to elementary science, Barr (1994) and Rowe (1993) raised the issue of equal access to science and conversation about science within classrooms, as did Cohen & Lotan (1991). They found that differences in status related to differences in access to materials and in opportunities to converse both in classroom and work groups. Low status students had less participatory opportunity. Rowe (1974b) reported differences in amount of wait time and pattern of verbal rewards related to gender and teacher expectations. Teachers need to guard against this kind of inadvertent inequity.

## SUMMARY

Principals are key to successful installation and maintenance of strong inquiry-oriented, activity-based science programs. They need guidance in two important categories: program logistics and program evaluation. Part of the instruction supervisors provide for administrators should include classroom visits in which they both help. Such visits provide a context for learning about program logistics and evaluation.

Teachers need to learn to distinguish between management of behavior and management of ideas as well as between conversation and inquisition. Training in wait-time techniques and group management skills is useful for these purposes. Science content background information for teachers should be presented as much as possible in contexts of interest to adults. Successful science programs depend on close collaboration of principals with teachers. Science supervisors can help this happen.

## References

Arendt, H. (1968) *Men in Dark Times*. New York: Harcourt Brace and World, 24.

Barr, B.B. (1994) Research on problem solving: elementary school. In *Handbook of research on science teaching and learning*, by D. L. Gabel (Ed.). New York: Macmillan, 237–247.

Cohen, E.G. (1984). The desegregated school: Problems in status, power and interethnic climate. In *Desegregation: Groups in contact: psychology of desegregation*, by N. Miller and M. B. Brewer (Eds.). New York: Academic Press.

Cohen, E.G., & De Avila, E. (1983) "Learning to think in math and science: Improving local education for minority children." *Final Report to the Johnson Foundation*. Stanford University, School of Education. Stanford, CA.

Cohen, E.G., & Intili, J.K. (1981 and 1982) "Interdependence and management in bilingual classrooms. Final report, NIE grant. Stanford: Center for Educational Research, Stanford University, Stanford, CA

Cohen, E.G., & Lotan, R.A. (1991), Producing equal-status interaction in the heterogeneous classroom. School of Education. Program for Complex Instruction. Stanford University. Stanford, CA

Johnson, D.W., & Johnson, R.T. (1987). *Learning together and alone: Cooperative, competitive, and individualistic learning*. Englewood Cliffs, NJ.: Prentice-Hall.

Nicholson, H.J., Weiss, F.L., & Campbell, P.B. (1994). In *Informal science learning, What the research says about television, science museums, and community-based projects*, by V. Crane, Ed., Chapter 4, 107–176. Research Communication, Ltd., Dedham, MA.

Rowe, M.B. (1974a). Relation of wait time and rewards to the development of language, logic and fate control: Part One–Wait Time. *Journal of Research in Science Teaching*. 11 (2), 81–94.

Rowe, M.B. (1974b) Wait time and rewards as instructional variable, their influence in language, logic, and fate control: Part II, rewards. *Journal of Research in Science Teaching* 11 (3), 291–308.

Rowe, M.B. (1986). Wait time: slowing down may be a way of speeding up! *Journal of Teacher Education*, January–February, 43–50.

Rowe, M.B. (1993). Science education, elementary schools. In *Encyclopedia of Educational Research, 6th ed.*, by M.C. Alkin, Ed., Macmillan, 1172–1177.

Rowe, M.B., McLeod, R., & Lawlor, F. (1995). Science Helper Video Series, developed at the University of Florida with funding from the National Science Foundation, (NSF #8751326MDR; NSF #9255703ESI). Armonk, N.Y., The Learning Team, Box 217, (in press).

## Author Note

**Mary B. Rowe** has been a teacher, lecturer, consultant, 1987–88 NSTA president, and professor of science education at both the University of Florida and currently at Stanford University. Her research on "wait time" in teaching has been considered one of the most significant discoveries in education in the past two decades. Her other major study, the role of early exposure to activity-based science programs in the development of fate control, has also been highly publicized. Her most recent project is the development of a middle school science program funded by NSF.

# Constructing Multicultural Science Classrooms: Quality Science For All Students

Mary M. Atwater
Denise Crockett
Wanda J. Kilpatrick

Policy makers, science leaders, science teachers, and community leaders are seeking research-based information, ideas, and suggestions, in addition to resources, that focus on providing quality science learning experiences for all K–12 students. Writers such as Atwater (1995b), Garcia (1988), and Kahle (1974, 1983, 1993) advocated the inclusion of all marginalized groups of students in learning quality science. Others such as Hill (Davis, 1994), Oakes (1990), and Shih (1988) championed specific groups of students to be included in the science learning dialogue. The American Association for the Advancement of Science (AAAS) (1989) identified the salient mathematical, science, and technological knowledge, skills, and attitudes that all students must possess at the completion of high school and established a conceptual basis for K–12 science education reform. The National Academy of Sciences (NAS) (1994) also has a draft of its standards for quality science learning and teaching. The purposes of these standards are to provide (1) a vision for science learning and teaching, criteria to judge the quality of what science students know and what science they are able to do, and (2) the standards for evaluating the programs that provide science instruction.

## DEFINITIONS AND DIMENSIONS OF MULTICULTURAL EDUCATION

In order to discuss "science for all students," certain terms must be utilized. The term Blacks will be used to describe citizens and other nationalities of African descent residing in the United States. Both the terms Hispanic and Latino will be utilized depending on the use by other referenced authors. Native Americans will refer to the indigenous people of the United States, Hawaiians, and Samoans. The term White will be used by these authors rather than European American or Anglo since neither of the two other terms is inclusive when referring to this ethnic group.

The following reasons provide a rationale for this chapter (Atwater 1995a). First, the demographics of K–12 science classrooms in this nation are changing. In 1970, students of color (Asian Americans, Blacks, Hispanics, Latinos, and Native Americans) and White students were 20 percent and 80 percent, respectively, of the K–12 school population (Simon & Grant, 1972). By the fall of 1991, the enrollment pattern has altered so that Asian Americans, Blacks, Hispanics, and Whites compose 3.4 percent, 16.4 percent, 11.8 percent, and 67.4 percent, respectively (Snyder, 1993). Presently, 31.6

percent of K–12 students are students of color. Secondly, a more science literate populace is needed in this nation if citizens are going to make informed decisions about the impact of science and technology in their lives. Employment in the future will greatly be dependent on the technological understanding of people, and more scientists and engineers are needed if the United States is to remain competitive in the global arena. Finally, if equity is to become a part of the science classroom and school climates, then all students must have the opportunity to learn quality science. This means that females, males, students of color, and students with disabilities will be learning quality science.

The five dimensions of multicultural education for K–12 classrooms serve as organizers for this chapter on science learning and teaching: knowledge construction, content integration, an equity pedagogy, prejudice reduction, and an empowering school culture (Banks, 1993a, 1993b). The dimension of prejudice reduction has been combined with equity in this chapter. Ideas about constructing empowered science classrooms are found in the last section. A listing of useful audiovisual resources is located in an appendix.

## CONSTRUCTING KNOWLEDGE FOR SCIENCE LEARNING

The science knowledge construction process defines the ways by which scientists construct knowledge and the manner in which the implicit cultural assumptions, frame of references, perspectives, and biases that influence the ways that scientific knowledge is created within the scientific discipline (Banks, 1995). Good (1995) believed that science is universal and maintained that science is the investigation of the underlying laws of nature. One of the premises undergirding *Science for All Americans* is that "scientists share certain basic beliefs and attitudes about what they do and how they view their work. These have to

do with the nature of the world and what can be learned about it" (AAAS, 1989, p. 25). Things and events in the universe occur consistently in ways that are understandable to humans through thoughtful, systematic investigations. Scientific ideas do alter as people's understanding of natural phenomena changes. These changes in ideas are usually modifications of ideas rather than rejections. From time to time, there are paradigm shifts in the thinking of scientists (Kuhn, 1970). Jane Goodall, Dian Fossey, and Birute Galdikas (the founding mothers of contemporary field primatology) altered the ways that scientists study primates and uncovered the significance of individual personality in primate groups (Morell, 1993). Many human concerns can not be examined by the scientific way; therefore, scientists are unable to provide complete answers to all of the questions people have. But what questions scientists can answer demand that evidence is available. Observations and measurements are rudiments for the evidence and provide information for the scientific arguments to make approximations about the world and how it works. Finally, these scientific arguments must explain and predict natural phenomena and be consistent with scientific observations.

Many people maintain that scientists try to avoid biases or strong inclination of the mind or preconceived opinion about something or someone. Good (1995) assumed that the laws of nature do not alter regardless of the many adaptations in the personal and cultural lives of scientists. Matthews (1993) stated the "world ultimately judges the adequacy of our accounts of it. How the natural world is is unrelated to human interest, culture, race, or sex" (p. 2). However, the underlying premises of science are preconceptions about what science is about and what should be studied in science. These assumptions limit what would be considered appropriate scientific questions and evidence.

Science literate students understand these scientific premises and assumptions. Science literacy is described as:

> understandings and habits of mind that enable citizens to grasp what those [scientific] enterprises are up to, to make some sense of how the natural and designed worlds work, to think critically and independently, to recognize and weigh alternative explanations of events and design trade-offs, and to deal sensibly with problems that involve evidence, numbers, patterns, logical arguments, and uncertainties (AAAS, 1993, p. XI).

The goal of school science as described in *The National Science Education Standards* is to prepare scientifically literate students capable of performing the following after 13 years of schooling:

> (1) use scientific principles and processes appropriately in making personal decisions; (2) experience the richness and excitement of knowing about and understanding the natural world; (3) increase their economic productivity; and (4) engage intelligently in public discourse and debate about matters of scientific and technological concerns(NAS, 1994, p. I–5).

These standards are based on the principle that "school science should reflect the intellectual tradition that characterizes the practice of contemporary science" (*National Science Education Standards*, 1996, p. 1–10). Others are now arguing for more inclusiveness in the frameworks for understanding the physical and biological worlds (Norman, 1995). In other words, a dialogue has begun about *what* science should be taught (Stanley & Brickhouse, 1994; Hodson, 1993). Multiculturalists, feminists, and philosophers of science have challenged the traditional thinking about the nature of science. Science has become problematic to these groups because its assumptions, preconceptions, and limitations reflect the culture and politics of the community of scientists who are mostly White, privileged males of Western heritage and are limited by their own vision of what is science (Stanley & Brickhouse, 1994; Harding, 1989; Hodson, 1993, Jegede, 1989).

Constructing meaning by science students requires that they actively seek to integrate new knowledge with knowledge already in their cognitive structures (Novak, 1993). Thus, science knowledge is understanding the nature of natural phenomena, acquired through a process in which students learn about the scientific values, goals, assumptions, and preconceptions. Students already have ideas about natural phenomena; therefore, their preconceptions can interfere with their scientific understanding. It is difficult to help students acquire another way of understanding natural phenomena. Consequently, science knowledge construction is being challenged within the science education community and by students everyday in science classrooms around the world.

## INTEGRATING CULTURE IN THE SCIENCE CURRICULUM

The more lenses through which students view natural phenomena, the greater their scientific understanding will be. Science teachers who use examples and content from a variety of cultures, groups, and their own personal experiences to illustrate science concepts, principles, and theories help make science more exciting and relevant to students' lives. "Too often we treat students as simply one more 'input' in a larger equation; yet as Dewey reminds us, our curricula always end in an act of personal knowing" (Apple, 1995, p. 136). If science teachers eliminate the personal aspects and the connections between the student and the school experience, then science curricula and teaching quickly lose both their vigor and their relevancy in the student's eyes (Apple, 1995). Many students fail when they sense no connection between the world of the school and their own individual and collective lives.

One of the authors, during her recent student teaching experience, perceived that many Black students were performing poorly in science. But within two months, many had improved their science grades because of the actions of the Black student teacher. She not only talked with them in class, but chatted with them in the hallways. Her actions demonstrated that she had a great interest in them as people and in their school work, especially science assignments. The science teacher might be the only science role model that students know. Atwater, Wiggins, and Gardner (1995) found a relationship between future science educational and career plans of urban students and their attitudes toward the teacher and the science curriculum. When science teachers connect with their students, students understand and value more science (Apple, 1995).

Apple (1995) asserted, "As educators have turned more to procedural models, to focus on 'how tos', we have paid much less attention to what we should teach and why we should teach" (p. 132). Apple insisted that the idea of a curriculum based on identifiable lists of names, dates, places, and things is quite limiting and does not provide a foundation for serious science education. The typology of multicultural curriculum that Baptiste (1993) has developed is based on the idea that educational concepts are divided into three levels of quality and quantity. At Level One, multicultural activities and ideas are inserted in the science curricula in merely an additive, superficial way. Designing a science program on ethnic celebrations such as Black History Month, Cinco de Mayo, and Women's Month is one example.

Level Two encompasses both product and process, unlike Level One, which focuses on only superficial products. Scientific contributions by other cultures, females, and scientists of color are integrated into the curriculum as opposed to being added. These science contributions are embedded in the curriculum, mirroring its pluralistic origin and development. Students learn that scientists from many cultures have made discoveries that expanded the body of science knowledge.

Level Three is the level at which multicultural science teachers operate. At this level, science teachers are involved as social activists. They help their science students promote equitable opportunity, have respect for those who are members of oppressed groups, and practice power equity in the school and the community. Furthermore, they help students to use their scientific knowledge to change the world around them. Herbal medicine provides ideas for inclusion in a Level Three curriculum that integrate curriculum and infuse culture. It incorporates science, health, social studies, and history. People of ancient civilizations of Africa, Greece, Rome, Japan, and Central and South America contributed to use of many of the medicinally valued plants today (McDaniel, 1994). Ancient Chinese widely utilized Ephredra (ephedrine) in treating asthma. Today, it is still found in many decongestants. The study of herbal medicines can lead to discussions about environmental issues such as the destruction of the tropical rain forest. With this approach, cultural and social issues are integrated into science learning.

Level Three science curriculum programs include a developed philosophy on multicultural science learning. The cultural experiences of Blacks, females, Hispanics, and Native Americans tend to promote a field-sensitive orientation that may partially explain these students' lack of success in science classrooms (Baptiste, 1993).

The characteristics of field-sensitive learners include the following:

1. in relation to peers, like to work with others to achieve a common goal, prefer to assist others, and are sensitive to feelings and opinions of others;

2. in the personal relationship to teachers, openly express positive feelings for teacher, ask questions about teacher's tastes and personal experiences;

3. in regard to the instructional relationship to teachers, seek guidance and demonstration from teacher, pursue rewards that strengthen relationship with teachers, and are highly motivated when working individually with teachers; and

4. in reference to a science curriculum, desire to have performance objectives and global aspects of curriculum carefully explained, want science concepts to be presented in humanized or story format, and like science concepts that are related to personal interests and other students' experiences.

For the field-independent science learners, the following characteristics apply:

1. in relationships to peers, prefer to work independently, like individual recognition, and are task oriented and inattentive to social environment when working;

2. in personal relationship to teachers, rarely seek physical contact with teachers, are formal and restrict interactions with teacher to tasks at hand;

3. in teacher-directed instructional activities, prefer to try new tasks and finish first, and pursue nonsocial rewards; and

4. in respect to the characteristics of curriculum that facilitate learning, prefer that details of science concepts to be emphasized

and parts of the science lesson to have meaning of their own, and prefer the discovery method for learning science.

A successful multicultural science curriculum includes both the integration of culture and the use of various teaching strategies to accommodate different ways of knowing.

## TEACHING SCIENCE EQUITABLY

I don't want to spend my time to listen to something I don't understand ... When my words come through my brain, and I couldn't, like have time for me to understand? And then, when I take the time to understand, then he [the teacher] is speaking another stuff (Harklau, 1994, p. 249).

"Teaching science is teaching students how to do science" (Lemke, 1993, p. xi).

Doing science in K–12 classes includes ways of talking, reasoning, observing, analyzing, and communicating what has been understood, and of using students' understandings to construct scientific arguments and interpret scientific findings. Language is an important element in science learning. Many science students are monolingual English speakers, others are monolingual speakers of a language different from English, and a few are bilingual or multilingual speakers. Science classrooms that are dominated by English speakers, including the teacher, become problematic for speakers of languages other than English. The English domination in science classrooms is sustained by systematically supporting the values and beliefs that support the superiority of standard English and establishes social relations in science classrooms that discourage the development of science literacy among students who are attempting to become bilingual (Darder, 1991).

McGinnis (1992) studied a school and science classes that supposedly valued diversity. In one teacher's class, McGinnis noticed that Mr. Green's English Speakers of Other Languages (ESOL) students sat together in the back of the center row of students. Throughout the study, the researcher never observed the ESOL students talking to other students in the class. Mr. Green gave the students the freedom to form their own cooperative learning groups, thus the students tended to be grouped together by ethnicity, language, and gender. He stated:

> It doesn't bother me … I think forcing it so you must have one Asian, one African American, one Middle Eastern, and one Caucasian in your group, I think that would be stupid. You know, I don't care. They all are friends (McGinnis, 1992, p. 213).

However, Mr. Green should have cared because the ESOL students did not know anything about the other students in the class and vice-versa. The English-speaking students did not even know the names of the ESOL students. Even the other students in the class remarked on the separateness and concluded that the ESOL students wanted to stay to themselves and did not want to associate with them. They were not friends, they were not even acquaintances. The ESOL students had little opportunity to share their scientific understanding in English and had no access to native English-speaking students to emulate or to assist them in improving their usage of English in explaining scientific concepts.

Science equity pedagogy prevails in classrooms when teachers have modified their instruction in ways that facilitate the science understanding of their students from different ethnic, cultural, gender, language, social class, and disability groups. With the rise of modern science and technology, social control has been exercised less through the use of physical discouragement and increasingly through the establishment of elaborate systems of norms and imperatives (Darder, 1991). Many White male students are encouraged to pursue advanced science and mathematics courses, while many females and students of color are tracked out of science (Kahle & Meece, 1994; Oakes, 1990).

Many K–8 science teachers exclusively employ cooperative learning and ignore other ways they can meet the needs of their students. Kahle and Meece (1994) concluded that cooperative learning "may not be effective in increasing girl's participation and achievement in science" (p. 550). Variety in instruction accommodates the diverse ways of student learning. In fact, diversification in science instruction meets the needs of students who learn by integrating information.

Teachers' knowledge about prejudice and of prejudice reduction is important for them to assist students' development of respect for differences in students. Teachers' and students' respect for differences helps create classroom environments where discussions about what students know and don't know can occur. These science classrooms become communities of science learning where all students can learn quality science.

## CREATING ENVIRONMENTS FOR EMPOWERED SCIENCE STUDENTS AND TEACHERS

In one junior high school I was invited to observe a history class by a teacher who admitted that he needed help with this particular group of students, all of whom were Latino. The teacher gave me a copy of his textbook, and I sat in the back of the room and followed the lesson … which was entitled 'The first people to settle Texas.' The teacher asked for someone to volunteer to read and no one responded. The teacher didn't ask for attention and started to read the text himself. It went something like this, 'The first people to settle Texas arrived from New England and the South in …' One hand shot up and that student blurted out, 'What are we, animals or something?' The teacher's response was, 'What does that have to do with the text?' (Kohl, 1992, p. 17).

The teacher stopped the lesson and introduced Kohl as the substitute teacher for the rest of the period. Kohl stood in the front of the classroom and reread the sentence from the book. Then he asked the students who believed this statement to raise their hands. A few of the students became alert. "This is lies, nonsense. In fact, I think the textbook is racist and an insult to everyone in this room," stated Kohl (1992, p. 17). All of the students woke up and one student asked the question, "You mean that? Well, there's more than that book that's racist around here" (Kohl, 1992, p. 17).

Power in classrooms is "maintained through selective silence and is manifested in the fragmentation of social definitions, management of information, and the subsequent shaping of popular attention, consent, belief, and trust" (Darder, 1991, p. 34). Empowered classrooms redistribute power among teachers and students so that both teachers and students become learners. Therefore, self-efficacy, student ownership of the classroom, and psychologically and physically safe environments become positive outcomes in these kinds of classrooms (Pate, McGinnis, & Homestead, 1995). The questions then become: How is an empowered science classroom established? What are the necessary elements for this kind of classroom? The following elements are needed for students to feel secure to ask questions, to express their opinion, and to view themselves as members of a community of learners (Apple & Beane, 1995):

1. The concern for all students by teachers and students;

2. An open dialogue that permits all students to be informed as possible;

3. The belief by students and teachers that students are able to solve problems;

4. The use of critical reflection to determine the value of scientific ideas, problems, and methodologies;

5. The school culture must promote democracy in all its programs and classes.

Dewey envisioned science as a way of knowing, democracy as a way of life, and science and democracy as foundations for a respect for diverse opinions (Hill, 1991). Dewey examined the inequalities perpetuated in classrooms and advocated for social change. He concluded that teachers can diminish the harshness of the social inequalities in schools and alter the conditions that create them (Apple, 1995). Science students become empowered when they are asked the following five questions (Meier & Schwartz, 1995):

1. How do you know what you know? (Evidence)

2. From what viewpoint is this interpretation being presented? (Perspective)

3. How is this event connected to others? (Connections)

4. What if the evidence was different? (Supposition)

5. Why are the evidence, information, and interpretations important? (Relevance)

The language and tools perhaps change in general science, biology, earth science, chemistry, and physics classrooms, but the questions remain the same to have empowered students.

Empowered teachers believe that the learning community within a classroom and school works for the collective good of all students. Banks (1993a) wrote:

> Grouping and labeling practices, sports participation, disproportionality in achievement, disproportionality in enrollment in gifted and special education programs, and the interaction of the staff and students across ethnic and racial lines are important variables that need to be examined in order to create a school culture that empowers students from diverse racial and ethnic groups and from both gender groups (p. 22).

To assist teachers in establishing science classrooms with empowered students, the authors compiled a list of audio-visual materials, found below. These materials provide different perspectives of scientific ideas, examples of effective science instruction for students from different cultures, and descriptions of some of the scientific contributions of women and people of color.

## Videos/Films

The Center for Applied Linguistics (Producer). (1990). *Communicative math and science teaching* [Video]. Washington, DC: The Media Center.

Covert, C. (Producer). (1992). *Gifted hands: The Ben Carson story* [Video]. Grand Rapids, MI: Zondervan Publishing House.

Ekulona, A.A. (Producer) & Garner, E. (Director). (1987). *The new ABC's: Preparing Black children for the 21st century* [Video]. Washington, DC: National Urban Coalition & S.M.A.R.T.

Englestad, K. & Warriver, G. (1996). *More than bows and arrows* [Video]. Seattle, WA: Camera One.

Harris, T. (Producer) & Edwards, J. (Director). (1989). *Crisis: Who will do science?* [Video]. New York: WNET.

Kessler, J.H. (Director). (1991). *Tracing the path: African American contributions to chemistry in the life sciences* [Video]. Washington, DC: American Chemical Society.

Phelan, K. & Sharpe, L. (Producer) & Phelan, K. (Director). (1991). *Winter wolf* [Video]. Seattle, WA: Miramar.

Quigley, B. (Director). (1991). *Teaching in the diverse classroom* [Video]. Seattle: University of Washington Video Production.

Rigg, M. (Producer) & Riggs, M. (Director). (1986). *Ethnic notions* [Video]. Mount San Francisco: California Newsreel.

Rooney, A.A. & Diamond, V. (Producer) & Wolf, P. (Director). 1957. *Black history: Lost, stolen, or strayed* [Video]. Xenon Entertainment Group.

Stoneburger, C.W. (Producer) & Stoneburger, C.W. (Director). (1991). *Women in science* [Video]. Madison, MI: Hawhill Associates.

## References

American Association for the Advancement of Science (1989). *Science for all Americans*. Washington, DC: Author.

American Association for the Advancement of Science, Project 2061 (1993). *Benchmarks for science literacy*. New York: Oxford University Press.

Apple, M. (1995). Facing reality. In *Toward a coherent curriculum*, by J. Beane (Ed.), p. 130-138. Alexandria, VA: Association for Superintendents and Curriculum Development.

Apple, M., & Beane, J. (1995). The case for democratic schools. In *Democratic Schools*, by M. Apple & J. Beane (Eds.), pp. 1–25. Alexandria, VA: Association for Supervision.

Atwater, M.M. (1995a). The multicultural science classroom. Part I. *The Science Teacher, 62*(2), 20–24.

Atwater, M.M. (1995b). The multicultural science classroom. Part II. Assisting all students with science acquisition. *The Science Teacher*, 62(4), 42–45.

Atwater, M.M., Wiggins, J., & Gardner, C.M. (1995). A study of urban middle school students with high and low attitudes toward science. *Journal of Research in Science Teaching*, 32, 665–677.

Banks, J.A. (1993a). Multicultural education: Characteristics and goals. In *Multicultural education: Issues and perspectives*, by J.A. Banks & C.A. McGee Banks (Eds.), pp. 3–28. Needham Heights, MA: Allyn and Bacon.

Banks, J.A. (1993b). Multicultural education: Development, dimensions, and challenges. *Phi Delta Kappan*, 75(1), 22–28.

Banks, J.A. (1995). Multicultural education: Historical development, dimensions, and practice. In *Handbook of research on multicultural education*, by J.A. Banks (Ed.), pp. 3–24. New York: Macmillan Publishing.

Baptiste, H.P., Jr. (1993). Multicultural education: Its meaning for science teachers. In *Science Matters*. New York: Macmillan/McGraw Hill.

Darder, A. (1991). *Culture and power in the classroom*. Westport, CT: Bergin & Garvey.

Davis, M. (Ed.). (1994). *Native Americans in the twentieth century: An encyclopedia*. New York: Garland Publishers.

Garcia, J. (1988). Minority participation in elementary science and mathematics. *Education and Society*, 1(3), 21–23.

Good, R. (1995, April). Taking science seriously. In *Can there be a Universal Science in our Multicultural World?* by N. Lederman (Chair). Symposium conducted at the meeting of the National Association for Research in Science Teaching, San Francisco, CA.

Harding, S. (1989). Women as creators of knowledge: New environments. *American Behaviorist Scientist*, 32, 700–707.

Harklau, L. (1994). ESL versus mainstream classes: Contrasting L2 learning environments. *TESOL Quarterly*, 28, 241–271.

Hill, P.J. (1991). Multi-culturalism: The critical philosophical and organizational issues. *Change*, 23(4), 38–47.

Hodson, D. (1993). In search of a rationale for multicultural science education. *Science Education*, 77, 685–711.

Jegede, O. (1989). Toward a philosophical basis for science education of the 1990s: An African viewpoint. In *The History and Philosophy of Science in Science Teaching: Proceedings of the First International Conference*, by D. E. Herget (Ed.), pp. 185–198. Tallahassee, FL: Florida State University, Science Education and Department of Philosophy.

Kahle, J.B. (1974). An alternative education experience: A workable program for science in large-city schools for the disadvantaged. *Science Teacher*, 41(9), 44–48.

Kahle, J.B. (1983). The myths of equality in science classroom. *Journal of Research in Science Teaching*, 20(2), 131–140.

Kahle, J.B. (1993). Ameliorating gender differences in attitudes about science: A cross-national study. *Journal of Science Education and Technology*, 2, 321–334.

Kahle, J.B. & Meece, J. (1994). Research on gender issues in the classroom. In *Handbook of research on science teaching and learning*, by D.L. Gabel (Ed.), pp. 542–557. New York: Macmillan Publishing Company.

Kohl, H. (1992, Autumn). I won't learn from you! Thoughts on the role of assent in learning. *Rethinking Schools*, 2, 16–17,19.

Kuhn, T. (1970). *The structure of scientific revolution*. Chicago: University of Chicago Press.

Lemke, J. (1993). *Talking science: Language, learning, and values*. Norwood, NJ: Ablex.

Matthews, M.R. (1993, April). *Multicultural science education: The contribution of history and philosophy of science.* Paper presented at the meeting of the National Association for Research in Science Teaching, Atlanta, GA.

McDaniel, P. (1994). *Role of plants in modern medicine: Structure, pharmacodynamics and use.* Unpublished manuscript. University of Georgia, Athens.

McGinnis, J.R. (1992). Science teacher decision-making in classrooms with cultural diversity (Doctoral dissertation, The University of Georgia, 1992). *Dissertation Abstracts International, 54*(2) 475A.

Meier, D. & Schwartz, P. (1995). Central Park East secondary school: The hard part is making it happen. In *Democratic Schools,* by M. Apple & J. Beane (Eds.), pp. 26–40. Alexandria, VA: Association for Supervision.

Morell, V. (1993). Called 'trimates', three bold women shaped their field. *Science, 260,* 420–425.

National Research Council (1996). *National science education standards.* Washington, DC: National Academy Press.

Norman, O. (1995, April). *Multicultural science education: Philosophical and historical questions of ownership.* Paper presented at the meeting of the National Association for Research in Science Teaching, San Francisco, CA.

Oakes, J. (1990). Opportunities, achievement, and choice: Women and minority students in science and mathematics. In *Review of research in education*: Vol.16, by C. B. Cazden (Ed.), pp. 153–221). Washington, DC: American Educational Research Association.

Novak, J.D. (1993, April). Meaningful learning: The essential factor for conceptual change in limited or inappropriate propositional hierarchies (LIPHs) leading to empowerment of learners. Paper presented as the opening lecture of the Third International Seminar on Misconceptions and Educational Strategies in Science and Mathematics, Cornell University, Ithaca, NY.

Pate, E., McGinnis, K., & Homestead, E. (1995). Creating coherence through curriculum interaction. In *Toward a coherent curriculum,* by J. Beane (Ed.), p. 62-70. Alexandria, VA: Association for Superintendents and Curriculum Development.

Shih, F.H. (1988). Asian-American students: The myth of a model minority. *Journal of College Science Teaching, 55,* 356–369.

Simon, K.A., & Grant, V. (1972). *Digest of educational statistics 1971 edition.* Washington, DC: US Government Printing Office.

Snyder, T.D. (1993). *Digest of Education Statistics 1993.* Washington, DC: U.S. Government Printing Office.

Stanley, W.B. & Brickhouse, N.W. (1994). Multiculturalism, universalism, and science education. *Science Education, 78*(4), 387–398.

## Author Notes

**Mary M. Atwater** is an associate professor at the University of Georgia, the author of numerous multicultural education publications, and a co-investigator of two Howard Hughes biomedical education grants worth $2.4 million. She has received several awards, including the 1990 NSTA OHAUS Award for Innovations in Four-year College Teaching for her efforts in multicultural education.

**Denise K. Crockett** is a doctoral student at the University of Georgia in the Department of Science Education. A former public school chemistry teacher, she received several local awards and grants. Her areas of interest include technology and cultural diversity.

**Wanda J. Kilpatrick** is currently a graduate student at the University of Georgia in the Department of Science Education. She recently completed a middle school science teaching internship and holds a bachelor degree in biology from Georgia College in Milledgeville, Ga.

# Collaboration: What Does It Mean?

Barbara S. Spector
Paschal N. Strong
James R. King

The authors describe a theoretical model for collaboration that emerged from two ongoing qualitative research studies. The model consists of 12 characteristics of collaboration, two factors influencing the expression of these characteristics, and a multifaceted continuum. To the extent that the 12 characteristics are all present in a given event involving individuals who are collaborating in science education, the collaboration is likely to be successful, and individuals are apt to experience the joy inherent in collaboration.

## COLLABORATION

*See that man in the big hat. He collaborates with the enemy. He passed classified secrets to Mr. Z. during the war.*

Collaboration means different things to different people.

The science education community has been experiencing a paradigm shift since the declaration of the science education crisis in 1982. The shift in science education is from the dominant reductionist approach to a holistic approach (see Figure 1). Inherent in the shift to a holistic paradigm is an awareness that the extant body of literature on leadership, group dynamics, group pro-

cesses, and problem solving does not seem to be enough to enable us to achieve the goals engendered in the holistic paradigm. Now educators are trying collaboration. What is there about the phenomenon of collaboration that has potential to move science education to where group process literature has not yet taken it? Educators need to understand how collaboration differs from other group initiatives, because the label collaboration is currently being used indiscriminately in the education reform movement.

The call for a multitude of collaborative initiatives is coming from various institutions, including government and funding agencies. But what does this word collaboration mean? Does it differ from cooperation? Is it the same as group process? Can educators be successful if we all come to the same collaborative initiative with different expectations and different understandings of the process and what can be achieved with it? Collaboration is an overused and underdefined word in the education enterprise.

## WHO IS COLLABORATING ABOUT WHAT?

For more than a decade, science educators have been called upon to collaborate

**Figure 1.** *Comparison of Characteristics of the Dominant and Holistic Paradigms.*

| DOMINANT PARADIGM | HOLISTIC PARADIGM |
|---|---|
| There is one objective reality independent of a person that can become known to an individual | Reality is constructed by individuals within their own minds. Therefore, there are multiple realities |
| Truth is correspondent to the objective reality | Truth is what a group working in a field at a given time agree to call reality (socially constructed) |
| The whole is equal to the sum of its parts | The whole is greater than the sum of its parts |
| Parts are discrete, each having their own identity | Pieces are altered when they interact to become part of the whole |
| Cause and effect are linear and immediate | Cause and effect relationships involve multiple factors, are complex, and may be difficult to distinguish |
| Hierarchies are the prevailing model organizing information, people, and things | Networks dominate the organization of information, people, and things |
| One can know the world by analyzing isolated smaller and smaller pieces | One can know the world by examining the whole |
| Science , using this reductionist approach, is the legitimate way of knowing | Science is one of several, equally valid ways of knowing |
| | The wholeness of the person, the union of the physical, spiritual, intellectual, and emotional aspects of an individual, is acknowledged |
| | Process is a product |

with stakeholders throughout the education and lay communities. Science supervisors and other leaders are commonly asked to initiate, organize, and maintain these collaborative initiatives, which vary in configuration. They may involve people from within the same discipline or across disciplines; within departments or across departments; between higher education and K–12 institutions; among elementary, middle and secondary schools; in business and industry or government agencies; and parents.

Perhaps the most prominent demand for collaboration is that which requires school science leaders to join with higher education faculty to work toward systemic change. This often means a school district leader must interact with faculty within higher education institutions who do not have experience collaborating effectively with each other. For example, higher education people do not normally work across departments, or among colleges within a university, or between four year and two year institutions.

The tasks on which science leaders are collaborating are as varied as the configurations. They cover everything that influences teaching and learning in formal and informal settings: teaching, reconceptualizing what teaching is, teacher enhancement, conducting research, program development for K–16, policy making and implementation. Collaboration has many faces.

## WHAT IS THE EFFECT OF RESPONSIBILITY FOR THESE COLLABORATIONS?

Having responsibility for such diverse initiatives in an enterprise that has not defined the meaning of collaboration can be exciting, or it can be overwhelming. The indiscriminate use of the label collaboration in science education has created a pitfall on our path to reform. People have come to collaborative initiatives with different expectations for (1) intended outcomes, (2) acceptance of responsibility, and (3) norms for behavior. This has, on more than one occasion, lead to disillusionment, frustration, and a loss of belief in the ability of the science education community to achieve meaningful change.

On the other hand, many science education leaders report deriving unprecedented pleasure and accomplishment from collaborations. The insights that can be derived from the theoretical model below have potential to help increase the number of people who experience the joy inherent in collaboration. The model is the authors' interpretation of preliminary findings from their two ongoing research projects on collaboration. The data base for one includes collaborations in science education for the past ten years. The data base for the second includes collaborations from other areas of education, from the sciences, from technology, from the arts, and from business and industry.

## WHY HAS THE DEMAND FOR COLLABORATION SURFACED?

Explicit in the holistic paradigm is the understanding that each person interprets the world through his or her own perspective, and that human beings must interact with each other in order to construct societal truth (Spector and Spooner, 1993). Thus it is necessary for science supervisors and other leaders to interact with representatives of all stakeholder groups to develop science education that can adequately serve our diverse population. An understanding of the multitude of perspectives held by the varied stakeholders in science education is essential if we are to ensure that all of us work toward common goals. That is why the demand for collaboration among stakeholders has surfaced.

## WHAT ARE SOME EXPECTATIONS AND PITFALLS?

If we are to engage in collaborations, we need to know what we are getting into. If someone is going to invest the enormous amount of time it takes to collaborate, he or she will surely want to know what rewards to anticipate. Will the effort bring pleasure? What are the banana peels one is likely to slip on? Can a person avoid any of the pitfalls?

A major pitfall stems from the diversity of expectations people bring to an initiative. When people come to something labeled collaboration, they come with varying expectations as to what it will be. The likelihood of success in a collaboration will depend on the extent to which these expectations are met. Expectations are usually a joint function of one's past experience in group situations and one's particular mode of interpersonal interaction. This suggests that educators collaborating to facilitate the paradigm shift must first spend time exploring each other's expectations of the collaborative process itself.

## WHAT ARE SOME CHARACTERISTICS OF COLLABORATION?

Twelve characteristics taken collectively differentiate collaboration from other human cooperative group activities. The order in which the characteristics are listed has no relation to importance in the collaborative process. The characteristics are: (1) equal empowerment, (2) a valued knowledge base, (3) trust, (4) commitment, (5) synergism, (6) emotional bookkeeping, (7) hedonic tone (including joy), (8) intrinsic motivation, (9) momentum, (10) time, (11) product, and (12) communication. The dyadic (two-person) group will be used as a model to elaborate on these characteristics.

To the extent that these characteristics exist in a group larger than a dyad, it is likely that the group will be more collaborative in nature and participants will be more apt to experience the joy inherent in collaboration. This paper addresses each characteristic individually. There is, however, considerable overlapping, because collaboration is consistent with the holistic paradigm where things are intimately woven into a complex whole that is greater than the sum of its parts.

### 1. Equal empowerment

The two individuals feel equally empowered and therefore no time is spent in power games or the establishment of superior-subordinate positions. This is true even when embedded in a more formal hierarchical structure. For example, there are reports of doctoral students and professors and K–12 teachers and professors working together on collaborative research projects in which each felt that all boundaries of academic rank disappeared during the collaboration. In some instances overcoming the culturally defined boundaries between roles happened naturally. In many instances, however, overcoming cultural expectations may require a deliberate effort.

### 2. Valued Knowledge Base

Each person perceives the other has a knowledge base different from his or her own and that the knowledge of the other is valuable, interesting, and would be useful to learn. There is a great deal of respect and admiration for each other's knowledge. The different knowledge bases are perceived to produce multiple perspectives on the task at hand. These perspectives often add disparate pieces to the puzzle. When all the disparate pieces are synthesized into a meaningful whole, there is an "AHA" experience.

### 3. Trust

An openness develops because of a feeling of trust and subsequent security. Ideas can be fully explored, developed, discarded, or altered because one has no fear of making mistakes, saying something foolish, or proposing a highly unorthodox proposition. This is the atmosphere that maximizes human creative potential.

### 4. Commitment

A commitment to make a collaboration work contributes to developing trust. Each person assumes a sense of responsibility for the other's well being. Commitment has at least two aspects: (a) People can depend on their collaborators to do what they say they will do, and (b) people can depend on each other for protection. This is not surprising when one realizes that a feeling of mutual commitment is important in any intense interpersonal relationship.

### 5. Synergism

The collaboration is synergistic. This means the participants become energized and the collaborative product is greater than the sum of its parts. In other group enterprises, such as committee work, the final product often represents a compromise between competing agendas and points of view. The final product may be a good, rational, achievable one. It may even be one that a member can take a certain amount of pride in, but it is unlikely to lead to startlingly new insights or results. In a collaborative relationship, on the other hand, one seldom feels the need to give something up to achieve something. There is no implicit or explicit bargaining going on.

## 6. Emotional Bookkeeping

One freely gives and receives without expectations of a strict quid pro quo relationship. There is no emotional bookkeeping going on. Regardless of actual time spent in the effort, each person gives because he or she wants to give and feels he or she is receiving full satisfaction in the relationship. There are no questions of who did more or made a greater investment. It would be extremely difficult to measure a particular contribution. For example, if one person has been writing for days and develops writer's block and the other just talks on the phone with the writer for an hour discussing a point that results in the writer unblocking, how can we measure its worth? If one person brings another into a network that took an entire career to build, how do we compare its worth to time invested?

## 7. Hedonic tone

Collaborations have a strong hedonic tone. The synergism of the collaborative process causes the process to be suffused with a strong hedonic tone. This strong affective component has two aspects: (a) The pleasure derived from the product of the collaboration, and (b) the pleasure of the interpersonal relationship involved in the collaboration. While the former is probably most always positive, the interpersonal component may range from highly, intimately involved to an active disliking. An example of the former is the collaboration of Masters and Johnson. An example of the latter is Gilbert and Sullivan. Thus these two affective components are somewhat independent of each other, although usually they are both positive.

Research by Mihalyi Csikszentmihalyi (1989) indicates that what people world wide label joy comes from an experience in which there is immediate and constant feedback. This feedback focuses one's attention on a task to the point where the person looses consciousness of self, looses self involvement, and is consumed by the experience. The rapid fire communication exchange in a collaboration can generate joy in this way. When both collaborators are immersed in the exchange and there is continuous feedback, a sense of union of mind and ideas, and shared goals, people experience joy. This is analogous to what Maslow (1970) means when he discusses the ability of the self actualized person to have a peak experience. People who are self actualized are most likely to collaborate well.

## 8. Intrinsic motivation

Humans work for two basic kinds of reinforcement: extrinsic rewards such as money, status, material possessions, or love and acceptance, and intrinsic reinforcements such as the pleasure in doing a particular job well, solving a difficult problem, creating something, or achieving a general sense of competency. While we all work for a combination of these types of rewards, maturity and a sense of real satisfaction from one's life seems to come more from intrinsic rather than extrinsic reinforcers. The collaborative relationship seems to tap these intrinsic reinforcers to a greater extent than many other types of human group endeavors.

## 9. Momentum

The collaborative process generates its own momentum. Its life does not depend on a particular goal or product. Instead, the collaborators continually generate new ideas and projects. It is as if the collaboration has attained a life of its own and is trying to find new justifications for continuing the collaboration. This does not imply that the *raison d'etre* for the collaboration is the collaboration per se, but the process is highly

---

reinforcing in and of itself. The process is also highly compatible with the human creative urge.

## 10. Time

Collaboration is independent of time. Other group cooperative activities, such as committee work, are often time constrained. They have a definite beginning, often a series of objectives to be met by a specific time, and often an end point. Many collaborations, on the other hand, just grow. Meetings are often unscheduled and spontaneous, and the collaboration continues after a particular goal or product is achieved. Collaborations may last for years or for a lifetime. When they do end, like a marriage, the reasons may be one of many and the parting may be anything from amicable to bitter.

## 11. Beyond product

Collaborations go beyond a product. This was noted before, but it is a significant enough factor in distinguishing collaboration from most group work to warrant a separate category. The collaboration is bigger than the specific goal, which may have been the original cause of the collaboration. While it has specific goals and purposes, the process itself is also a strong source of motivation.

## 12. Communication

People in a collaboration explicitly question each other's understandings of labels used to communicate. They are free to argue for other interpretations and labels. They have the liberty to appropriate and change interpretations of each others words and thoughts. There is fast give and take which individuals perceive as intriguing, rather than threatening. The skill of a phenomenologist, one who attempts to walk in another's shoes and report the other's reality, is used as a tool to question each other's meanings.

## WHAT FACTORS INFLUENCE THE PRECEDING CHARACTERISTICS OF COLLABORATION?

Two factors influence the expression of the preceding characteristics of collaboration: personal contact and small size effects.

### A. Personal contact

Collaborative relationships seem to require a minimum amount of time together or propinquity. It could be likened to a love affair. Love letters are nice but hardly a substitute for the real thing. There are collaborative relationships that attempted to continue after the partners had to separate geographically, but it is common for them to gradually dissipate and eventually end.

### B. Small size effects

Collaborative relationships become more difficult as the size of the collaborative group increases. Think of a group of three individuals in a collaborative relationship. Humans in groups of three have certain tendencies in alliance formation that can be detrimental to the collaborative process. The possibility of schismatic forces increases as the group grows larger.

## WHAT IS A CONTINUUM FOR THINKING ABOUT COLLABORATION?

It is useful to think of the varied configurations for collaborative initiatives in which we are asked to engage as being on a continuum. One end of a continuum is a dyad. The other end can be labeled the highly structured, vertically hierarchical group. So far as a group initiative fits the preceding theoretical model, it approaches an ideal of collaboration. As a particular group effort deviates from the theoretical model, it moves away from this concept of collaboration. Unfortunately, the majority of educators are being asked to establish collaborations within the hierarchical structures of most current educational organizations, even though collaboration is consistent with the holistic paradigm, not the reductionist paradigm. The hierachical structures are antithetical to collaboration. The continuum for collaboration is multidemensional. Here are six dimensions that comprise the continuum:

## 1. The size dimension

It has already been suggested that collaboration becomes more difficult as the size of the group increases, probably as a function of the square of the number. For groups working cooperatively on a specific problem, such as an ad hoc committee, group effectiveness first increases with size and then decreases after a critical size is reached.

## 2. Equal empowerment dimension

A dyad can consist of two equally empowered individuals, but it may not. Some individuals may have personal needs, such as power or dependency, which can make it difficult or impossible to enter into an equal relationship. As groups become larger, the probability of certain individuals taking charge or letting others take major responsibility for the work makes the equal empowerment component more difficult to manage. Yet, it is possible for such a condition to exist in fairly large groups, such as in academic departments in a school or a university.

Other groups, however, may be highly structured with a clear cut vertical chain of command from the outset. Such groups have a tendency to be adverse to building trust and open communication, and various power games ensue.

A special case of collaboration can exist when people come together because they feel impotent individually. The motivation to collaborate is to gain power as a group. These people may see the collaboration process as a mechanism for empowerment.

## 3. Time and goal dependency

There is movement away from the conditions that encourage creativity and spontaneity—the essence of the collaborative relationship—as a group becomes more focused on achieving a specific goal within a rigid time frame.

## 4. Propinquity

As groups from different departments, institutions, or geographic areas are formed, the physical distance increases among collaborators. The greater the physical distance, the harder it becomes to develop the kind of physical contacts necessary for collaboration. In so far as new, interactive communication technologies are developed, these effects may be mitigated to some extent.

## 5. Depth of Involvement

Even in a dyadic group meeting all of the other characteristics discussed thus far, there may be difference in the depth of participants' involvement in the collaboration. The involvement may vary from relatively superficial to deeply intense. There are different degrees of emotional attachment to the collaborative relationship. Even in somewhat larger groups, one may be deeply committed or more superficially committed.

## 6. Culture

People respond to their perceptions of reality, not to some objective reality. A person's world view, which commonly emerges from culture, is therefore the lens through which one will interpret the interactions in a group. Culture here refers to ways of doing things that signify a particular group of people and their products, art, and language. The culture from which members of a group have come influences the expectations a person has for acceptable behavior in a group. The degree of congruence within a group in terms of shared values, beliefs, similarity of experiences, and ethos, will influence the extent to which collaboration develops. This congruence does not imply a rigid conforming "group think." It is a congruence of basic values that allow for the tensions, dissonance, and ambiguities that drive the creative process. In other words, people agree to disagree.

## WHAT IS THE IMPACT OF COLLABORATION IN SCIENCE EDUCATION?

It takes all the characteristics listed here, in combination, to create a holistic view of the unique human activity called collaboration. To the extent that these characteristics exist in a group, a group is establishing a collaboration and is likely to derive the satisfaction inherent in the process.

Science leaders can help to enable all our colleagues to share in the joys from the learning, the interactions in the relationship, the sense of security of having colleagues you can count on, having someone to share the weight of a task, and the elation of creating a product no one has previously made work or even envisioned. By helping people examine their expectations and explicitly share those expectations, the joy people derive from collaboration can be increased. As we analyze a particular group effort and wish it to be truly collaborative, we must be aware of many factors and consciously address them. We must also develop strategies that will be effective in educating our future science teachers and all science leaders to be good collaborators.

## References

Csikszentmihalyi, M. (1989). *Learning joy, and survival ideas for a creative curriculum*. Paper presented at the 44th Annual Conference of Association for Supervision and Curriculum Development. Orlando

Maslow, A.H. (1970). *Motivation and personality* (second edition). New York: Harper and Row.

Spector, B. & Spooner, W. (1993). The changing role of the science supervisor: A response to a changing paradigm. *Sourcebook for science educators, fourth edition*. Washington, DC: National Science Teachers Association and National Science Supervisors Association.

## Author Note

**Barbara S. Spector** is a professor of science education at the University of South Florida where she directs the Science/Technology/Society Center, an umbrella for multiple innovative programs implementing a vision of the schools of tomorrow and the Project 2061 Higher Education Research and Development Center. She has been awarded 43 grants supporting the restructuring of science teacher education. She has authored more than 300 books, chapters, articles, and presentations and served as consultant to state and local education agencies and professional associations. She was research director for the National Science Teachers Association and has served on the Board of Directors of several other national science education associations.

**Paschal N. Strong** is a full professor in the Psychology Department at the University of South Florida. He has been at USF for 30 years. His areas of specialties include neuropsychology and comparative psychology. He has thirty publications, five book chapters, and several research grants. He has worked in the areas of Hippocampus functions, Alzheimer's disease, and comparative intelligence of animal species, including primates, up to chimpanzees. He has also been a part of the laboratories involved in training chimpanzees for the space shot for NASA.

**James R. King** is an associate professor of reading education at the University of South Florida. His research is in feminist and critical literacies, in emergent and at-risk literacy, and in qualitative research methods. After completing graduate work at West Virginia, Dr. King taught at the University of Pittsburgh and Texas Woman's University. In 1990, Dr. King spent a sabbatical teaching first graders.

---

# Transforming Science Education In Ways That Work: Science Reform In The Elementary School

James A. Shymansky

Over thirty years ago, the post-Sputnik reform of science education was in full bloom, and there was anticipation of positive changes in what and how students would learn science. There was particularly a sense that science education in the elementary grades would change radically to enable students to think and do science in engaging and meaningful ways. Despite the great support for and teacher participation in many of the new alphabet science curricula of the era, the post-Sputnik reform sputtered by 1970 and gave way to a reinvigorated "back to the basics" movement.

The current science education reform, sparked by Project 2061 and articulated in *Science for All Americans* (SFAA) (Rutherford & Ahlgren, 1990) and *Benchmarks for Science Literacy* (Project 2061, 1993), challenges us to avoid the pitfalls of the post-Sputnik era and to use what was learned to ensure that all students have the opportunity to know, understand, and do science in developmentally appropriate ways.

"SFAA answers the question of what constitutes science literacy" while "*Benchmarks* specifies how students should progress toward science literacy" (*Benchmarks*, p. XI). These publications and the broader Project 2061

effort chart a course that challenges all players—teachers, students, parents, school administrators, school boards, and the larger community—to rethink their views of what it means to teach and learn science. As we rethink what science education can and should be, as well as what we come to recognize in students as evidence of knowing, understanding, and doing science, it is clear that reform is a journey rather than a state to be achieved. And like any journey, the reform course charted by Project 2061 must be viewed through the lens of the cultures of the schools, communities, families, teachers, and administrators across this nation.

As we reflect upon our lost opportunities in past reforms and as we learn more about how children learn, it is evident that enduring change comes only when the culture of schools supports reform (Sarason, 1990, p. 130). The purpose of this article is to examine the culture of the elementary school in order to understand and propose ideas and methods to facilitate the growth and development of science literacy in young children.

## THE ELEMENTARY SCHOOL

If current reform in science education is to have a sustained impact, we must address what special features of elementary school culture we should consider as we seek to promote science literacy in young children. The following features have special relevance to the reform challenge. Imagine a teaching assignment of ten or more preparations. Imagine a class rich in diversity—children varying in backgrounds, interests, and abilities. Imagine a classroom with limited materials beyond printed texts and paper and pencils. Then imagine having major responsibility for the emotional, social, intellectual, conceptual, and motor skill development of these children. Add to these responsibilities playground, lavatory, and lunchroom duty, and a picture of what it's like to be an elementary school teacher begins to take form. In addition to teaching all the curricular areas, an elementary school teacher has to know about child development, nutrition, behavior disorders, and family counseling and know how to fix broken zippers, untie knots, play kickball, and eat lunch in ten minutes.

## Broader Responsibilities

While all teachers experience some of these conditions and problems, elementary teachers are unique in their responsibilities. To begin, an elementary school teacher is often responsible for ten or more instructional areas (Goodlad, 1984). Science is but one of many areas that make up this curriculum—and one with a very low priority in the grand scheme of things at that. The K–6 curriculum targets writing, reading, and spelling as well as traditional areas such science, mathematics, geography, history, art, foreign language, music, and health as formal areas of instruction. The K–6 teacher must be somewhat expert in all of these. Most middle and high school teachers, in comparison, have the luxury of being able to specialize in one or two content areas.

## Curriculum Organization and Planning

Another major distinction in the elementary school is the integrative nature of the curriculum. Because the curriculum is so chock-full, K–6 teachers routinely teach "across the curriculum." They still plan and teach within single curricular areas (e.g., reading, math, spelling), but they frequently plan instruction around broad themes (e.g., black history month, fire prevention, rain forests) to focus activities in multiple curricular areas including science. In many classrooms, the only science that gets taught is that which somehow fits into a larger curricular theme.

Curriculum planning is also unique to the elementary school. There are usually no departments per se. Formal planning is done at the building level and varies with the leadership of the principal. There is almost no program articulation with the middle and high schools. The science curriculum is often determined by the district textbook program or left to the discretion of the individual teachers, who often simply split up the program and pick out favorite units. Few K–6 teachers have experienced teaching a science program that is organized conceptually and developmentally.

## The Family Factor

Another distinguishing characteristic of the elementary school setting is the close tie with the parent community. Parents of K–6 students have historically taken a more active role in their children's schooling than have parents at the upper grade levels. They spend more time in the classroom assisting teachers and are more involved with after-school projects and homework (Fullan & Stiegelbauer, 1991; Swick, 1991). Schools are increasingly being called upon to assume greater responsibility for before- and after-school activities as the incidence of working single-parent and two-parent families continues to rise. The growing pressure to extend the school year and add new areas of responsibility to the schools results in a curriculum that is in a constant state of flux. And the problem is no more pronounced than at the elementary school level, where the school has been historically viewed as an extension of or substitute for a nurturing home environment.

## IMPLEMENTING THE BENCHMARKS

With these features common to elementary schools across this country, how can the vision expressed in *SFAA* and laid out in the *Benchmarks* be implemented and sustained in the context of this elementary school culture? Several of the entries in the *Benchmarks'* Chapter 14, "Issues and Language," seem to offer a particularly good platform for discussion: (a) Knowledge and Knowing, (b) Grain Size, (c) Connections, (d) Curriculum Blocks, (e) Habits of Mind, (f) Interdisciplinary, (g) Less Is More. Let's examine each of these and explore their implications for the elementary school culture.

### Teaching and Learning for Understanding

Developers of the *Benchmarks* opted for the simple words "know" and "know how" and chose to use "plain English" in framing the *Benchmarks* to avoid the ambiguity of action-oriented verbs and to discourage the premature and inappropriate use of technical language (p. 312). What these decisions mean for instruction, curriculum development, and assessment is not exactly clear by the authors' own admission (p. 320). The intent, however, is clear: developers wanted to promote learning for understanding over rote memorization. Perkins and Blythe (1994) define understanding as "being able to carry out a variety of 'performances' that show one's understanding of a topic and, at the same time, advance it" (p. 6). In the *Benchmarks* it is added "that what counts is lasting knowledge and skills, not just what one would know or be able to do *at the moment* of completing school or reaching any particular grade level" (p. 320).

Perhaps of all the areas of research related to the *Benchmarks*, the issue of what constitutes understanding is the most critical. It will be the most difficult to address, because most teachers think they already teach for understanding despite the growing body of literature that argues to the contrary. It is one thing to set understanding and the development of critical response skills and other habits of mind as goals; it is altogether something else to promote and facilitate them (to teach for them) and to know when a student has developed them (to assess them). For K–6 teachers, teaching and learning for understanding is a very confusing issue because teachers themselves have so little understanding of science as a way of knowing and thinking. Since so much of their own learning was of the rote memorization type, teachers are ill prepared to teach for understanding. Science to most is a body of knowledge to be transmitted.

Simply infusing hands-on activity into the curriculum, without a major adjustment in how teachers view teaching and learning, is not the solution either. Three decades of hands-on science have not produced any dramatic improvement in the understanding or habits of minds of students or teachers, as various national and international surveys attest.

So one of the first orders of business suggested by the *Benchmarks* is to have teachers grapple with what it means to understand something in science. Perhaps the best way to address the question is through assessment reform. Rather than seeking new ways to teach science for understanding as the first step in reform, it may be more successful and satisfying to start by describing the outcome and how it will be assessed (Champagne, 1994).

The current flurry of interest in performance assessments and portfolios suggests that the larger education community may be ready to tackle problems of teaching and testing for understanding and problem solving, but these strategies are difficult and costly to develop. Following the maxim that "good assessment activities make good teaching activities and vice versa" (Shavelson, Baxter, & Pine, 1991, p. 350), developers have drawn heavily on popular instructional activities in the initial rounds of performance test construction. But a word of caution seems prudent: cookbook performance tasks are probably no better than multiple-choice tests at finding out what students understand or how they are thinking.

## Less is More

Perhaps no words are more associated with the current reform in science education than the phrase "less is more" (LIM). The phrase and its implications for the classroom are also the least understood. Tied to the LIM issue are the notions of big and important ideas of science and grain size, which indicates the extent to which concepts are grouped into related sets. The LIM theme is particularly relevant to K–6 instruction. It indirectly addresses the question of what to teach. When K–6 teachers teach from textbooks, this question is substantially answered for them (however inappropriately). But now, most new programs are using a stand-alone-module format, and teachers are being encouraged to mix and match materials and activities. As teachers make decisions about what to teach and when to teach it and as curriculum developers, assessment teams, and teacher educators try to evaluate their current practices and implement changes, attention will center on how the LIM theme translates into what students should be able to do and understand in science.

Conversations and interactions with many different teachers reveal a fairly common interpretation of the LIM phrase: "more depth and less breadth." Phrases such as "quality over quantity," "understanding over memorization," and "big ideas over trivial facts" are also heard. What is not very well understood is what these phrases mean at the planning, teaching, or testing level. Two problems of interpretation have particular potential for thwarting a substantive reform effort or sending it in the wrong direction altogether. There is danger that LIM could be translated literally and used to reduce the time spent doing science. This is especially problematic at the elementary school level, where many teachers already look for reasons not to do science. There is also the danger that LIM will be used to justify teaching traditional science courses (e.g., biology, geology, chemistry) at earlier grades to allow for depth of study—a move that would clearly be contrary to the spirit of the LIM theme.

There are several important questions raised by the LIM theme. The first thing K–6 teachers will want to know is what are the important, big ideas on which their science instruction should focus and what ideas and topics that are currently included in their program of instruction should they deemphasize or eliminate? This is an issue that is skirted in the *Benchmarks* with phrases such as, "*Benchmarks* does not advocate any particular curriculum design" (p. XII). Even the descriptions of curriculum blocks, which will be discussed in Project 2061's to-be-developed *Designs for Science Literacy* (referred to in the *Benchmarks*, p. 381), will not provide the complete answer teachers seek. It is very important then for teachers at the building or district level to go through the *Benchmarks* and decide a scope and sequence that fits.

## Integrated Science

In the *Benchmarks*, the emphasis in LIM is on the connectedness and coherence of knowledge across science, mathematics, and technology (p. 320). There is discussion of concepts increasing in sophistication across grade levels (spiraling) and suggestion that two or more concepts may connect with other concepts or converge to form new, more complex concepts at subsequent grade levels (p. 315). It is recommended that ideas cut across disciplines and span grade levels using "strands" or "story lines" to build the connections (p. 315). This recommendation plays well in the elementary school where teachers are used to using story platforms and organizing around themes (Butzow & Butzow, 1988). But integration and teaching across the curriculum have a special meaning for K–6 teachers. Thematic instruction traditionally cuts a swath that is much broader than science, mathematics, and technology. The language arts are emphasized more than the other areas combined. This predisposition of K–6 teachers to stress language arts in the curriculum could take the science instruction in radically different directions. If the story line and strand ideas in science can be shown to enhance instruction in the more emphasized language arts and achieve the science literacy goals of Project 2061, a real coup will have been accomplished. Early in the reform effort, teachers should explore ways to teach general reading, writing, and listening skills in the context of hands-on science activities through the use of special science literature, storytelling, drama, and visual arts.

## A Move to Whole Science Instruction

Practically every K–6 teacher in the country is familiar with the whole language movement—using real books to teach reading literacy in a natural context (Tunnell & Jacobs, 1989). In a sense, Project 2061 represents an effort to promote a whole science movement. The whole language movement attacks the basal-text, bottom-up, reading-skills approach. In a parallel manner, Project 2061 attacks the rote memorization accumulation of science facts and promotes the understanding of science as a way of thinking and learning within an ever-expanding set of interconnected ideas.

Whole language instruction uses real books. Everyone knows what real books are; they are books read for enjoyment or information. The analog in whole science would be real science. But what is real science? Post-Sputnik curriculum reformers defined real science as that which scientists do. It was a basic tenet of 1960's reform that science would be intrinsically interesting and meaningful to students if learned in ways that scientists experience it. Is that idea again the essence of Project 2061? Actually not. More important than the products or the processes of science are the habits of mind that students develop as they learn and do science. These are the "values, attitudes, and skills ... [that] relate directly to a person's outlook on knowledge and learning and ways of thinking and acting" (*Benchmarks*, p. 281).

There are other parallels between a whole language and whole science movement that are worth developing in the K–6 arena. In whole language instruction, the learner pursues areas of personal interest; the content of the reading is left wide open. *Benchmarks* sets goals that define science literacy but leave unspecified the paths to be taken in pursuit of those goals. An instructional strategy that has received considerable attention in recent years connects science, technology, and societal issues (STS). In an STS approach, science learned in contexts that are personally and societally relevant is learned with greater depth and understanding (Yager, 1993).

But can a more open-ended curriculum work in the elementary school, where teachers are less confident in their science expertise than their counterparts in the upper grades and where children are less experienced and worldly? If children are left to their own interests and allowed to pursue more personalized topics of study, will they learn the important ideas and develop the habits of mind described in the *Benchmarks*? Certainly individual K–6 students have been known to strike out on their own to investigate topics and questions in science with great learning success. But can such a whole science program work with a K–6 classroom of 20–30 students and teachers who are unprepared in science?

A whole science movement probably won't work under existing educational models that view the classroom teacher as the major if not sole source of instruction. Whole science instruction will require a redefinition of what it means to teach and who can be considered a teacher. The larger school community of teachers, parents, and business partners will be needed to promote and develop in students the science literacy called for in Project 2061 (as in the old African proverb, "the whole village educates the child").

The notions of whole science instruction specifically, and Project 2061 generally, raise the more fundamental question of how to change individual and community paradigms. Rote memorization of the planets in the solar system or the parts of a flower is a practice that will not die easily among K–6 teachers, parents, or students. Effective change and implementation will have to involve the whole community. And the process is likely to take several years (Hall, 1979; Hand, 1992).

## A NATURAL NEXT STEP

If we are to transform elementary science education in ways that work and endure, we must build upon what is known about pedagogy, learning theory, and the strengths of K–6 teachers. Whole science instruction does just that. The use of language to promote inquiry, analysis, application, and reflection in science is a logical and powerful strategy for elementary school teachers. Whole science instruction does not compete with reading and language arts for time in the busy elementary school curriculum; it complements them. More importantly, it builds on and respects elementary teachers' intuitive and practiced skills in engaging and challenging students in learning. Whole science instruction encourages and supports

thinking and doing science as an extension of purposeful reading, listening, and writing. It builds upon the use of language in its richest sense to make science content interesting, important, and meaningful for students and teachers.

Whole science instruction also directly supports assessment reform. New forms of assessment, such as portfolio assessment, performance based assessment, and other alternative ways to systematically document what students think, understand, and can do in science, readily become extension activities for whole science instruction. The natural next step for elementary school teachers is to meld the pedagogy of whole science instruction into a seamless web of teaching, learning, and assessment. The connections among teaching practices, learning experiences, and assessment opportunities bring power and credibility to the logical argument to use language to promote science learning.

## References

Butzow, C.M., & Butzow, J.W. (1988, February). *Science, technology and society as experienced through children's literature.* Paper presented at the Science, Technology, Society Conference on Technological Literacy. Arlington, VA.

Champagne, A.B. (1994). [Interview]. In *Assessment alternatives: Finding out what students know and can do* [video]. Washington, DC: Council of Chief State School Officers.

Fullan, M.G., & Stiegelbauer, S. (1991). *The new meaning of educational change.* New York: Teachers College Press.

Goodlad, J.I. (1984). *A place called school: Prospects for the future.* New York: McGraw Hill.

Hall, G.E. (1979). The concerns based approach to facilitating change. *Educational Horizons, 57,* 202–208.

Hand, B. (1992). *Constructivist approaches to teaching and learning in a secondary school science depart-*

*ment.* Doctoral dissertation, Curtin University of Technology, Perth, Western Australia.

Perkins, D., & Blythe, T. (1994). Putting understanding up front. *Educational Leadership, 51*(5), 4–10.

Project 2061, American Association for the Advancement of Science. (1993). *Benchmarks for Science Literacy.* New York: Oxford University Press.

Rutherford, F.J., & Ahlgren, A. (1990). *Science for all Americans.* New York: Oxford University Press.

Sarason, S.B. (1990). *The predictable failure of educational reform: Can we change course before it's too late?* San Francisco: Jossey-Bass.

Shavelson, R.J., Baxter, G.P., & Pine, J. (1991). Performance assessment in science. *Applied Measurement in Education, 4*(4), 347–362.

Swick, K. (1991). *First: A rural teacher-parent partnership for school success.* Final report to the U.S. Office of Education (First Division), Columbia, SC.

Tunnell, M.O., & Jacobs, J.S. (1989). Using "real" books: Research findings on literature based reading instruction. *The Reading Teacher, 42,* 470–477.

Yager, R.E., (Ed.) (1993). *The science, technology, society movement.* (What research says to the science teacher, 7). Washington, DC: National Science Teachers Association.

## Author Note

**James A. Shymansky** is a professor of science education at the University of Iowa. Author of numerous journal articles, chapters, and textbooks, he has also written and directed many science education grants. Honors include AETS Outstanding Science Educator, NSTA Gustav Ohaus Award for Innovations in College Science Teaching, and the University of Iowa Excellence in Teaching Award.

# Staff Development Issues for the Science Leader

## Don H. Kellogg

Politics and education are indeed strange, and many would say inappropriate, bedfellows. The result of one such liaison, the National Education Goals (Goals 2000: Educate America Act, 1994), has been nothing if not controversial. Goal five of Goals 2000, which states that by the year 2000 "U.S. students will be first in the world in math and science achievement", is one that has generated its share of controversy among science educators. One reason for the controversy is that many teachers and scholars see this goal as being ill designed. They believe, perhaps with justification, that there is no valid set of international standards by which "first in the world" can be determined.

However, within these eight goals is one that generates considerable support among educators and, if achieved, is certain to bring about significant improvement in science education and in all other areas of the curriculum. That goal is number four: "the nation's teaching force will have access to programs for the continued improvement of their professional skills and the opportunity to acquire the knowledge and skills needed to instruct and prepare all American students for the next century."

Whether it is known as inservice education, professional development, or staff development, the continued professional growth of teachers is and has been a stepchild in the overall teacher education scheme. When Yogi Berra said, "It's not over till it's over!" was he talking about teacher education? Is there a strong tendency to see that last degree earned, be it bachelors, master, or doctorate, as being the end of teacher education? Certainly most states have relatively specific requirements for preservice teacher education, but less stringent requirements are imposed on inservice teacher education. States and/or school systems often require X hours of staff development within Y years, but usually nothing is said about the content or chronology of these experiences. The requirements can often be satisfied by college courses (no matter what the content), by attendance at professional meetings, and by the infamous teacher inservice programs conducted in that week before the students return to the classrooms after summer vacation.

This approach has certainly created a windfall for the entrepreneur. The aggressive and ambitious university faculty member has only to write a proposal that is fundable, then recruit some teachers and hold a summer workshop. Or those entrepreneurial faculty could take the "hit and split" approach of conducting two-hour seminars, for profit, in nearby school systems. In no way do I mean to imply that all staff development activities of these types were of no value. Many were exceptional, even watershed experiences for the participants, but far too many suffered from at least one of the three limiting characteristics that follow.

1. *They reached a relatively small percentage of science teachers.* This was and is true especially of projects in which elementary teachers participate. Most funded projects that include summer workshops imposed on participating teachers an obligation to conduct local inservice to gain a large multiplier effect. This approach may have reached a significant percentage of teachers in a given system, but in the overall picture, the percentage of teachers reached was very small.

2. *There was little or no coordination among various staff development projects.* The fact that these projects were only peripheral to any greater coordinated effort was perhaps their greatest weakness. In spite of the rhetoric about systemic change and national goals, professional development planners at the local schools, the state departments of education, the universities, and frequently the national entities have yet to develop a degree of coordination that makes efforts united in either actuality or appearance.

3. *They were of relatively short duration with little or no extended follow-up and support.* This has been almost universally true of the "hit and split" inservice, and far too many summer workshop type projects have provided little more than token follow-up support. Change in the classroom rarely comes about in a rapid and linear manner; it takes considerable time, a good deal of effort, and it is usually accompanied by a few failures and false starts along the way. If encouragement

and support are not forthcoming during the inevitable setbacks then the most common outcome is reversion to the status quo.

An additional issue related to staff development is that staff development efforts are often perceived by teachers and even some science leaders as remedial. For over 30 years massive amounts of money have been spent by the National Science Foundation, Department of Education, Department of Energy, and a host of others to support staff development activities that have at their core changed the way science teachers teach. For the teacher in the trenches the implied message is that the way they are teaching is wrong or that "someone" just wants them to teach differently.

The ensuing result is that staff development activities with teachers who have made either of these inferences are usually exercises in futility. Teachers are notoriously independent and are not above resisting external pressure for change unless they are convinced the change is an improvement over current practice. Experienced teachers have often been through many cycles of "innovation" and often demonstrate considerable skepticism about what they may see as just the most recent fad. This is in no way an indictment of teachers but is rather a recognition of human nature. As a result, in 30 years, the billions of dollars spent and the untold number of staff development sessions have not significantly changed teaching or learning in the average science classroom.

So how might some significant improvement be made in inservice science education? A first step might be to address the absence of systematic and coordinated professional development plans and activities for systems. Most states have science goals or frameworks, and the majority have recently been updated to reflect the latest concerns and trends. Those frameworks or policy statements usually recommend professional development for teachers, and this is frequently all the attention given to the issue.

To build upon policy statements, needs assessments must be conducted to determine what types of professional development activities should get attention at each of the local, regional, and state levels. Of all the needs assessment that may be conducted, none is more important than the local, and it must involve all teachers who teach science if it is to be of value in planning and conducting professional development. The results of these needs assessments should be studied carefully and made a part of the plans for both local and state programs such as the Dwight D. Eisenhower programs.

Planners of local professional development activities should be aware of statewide needs assessment and state professional development plans but must consider them in light of local needs. In some cases local needs and state plans may not be congruent, and local needs should usually be met before state professional development plans are implemented.

One factor that contributes to these comprehensive and collaborative planning activities not being accomplished is the long standing and often bitter turf struggles going on within a state. Players include higher education, public schools, and state education agencies. In order to achieve coordinated systemic professional development plans those struggles must be overcome and the tasks approached as a unified team. Sadly we frequently find this objective very difficult to accomplish. If we were able to put aside these numerous differences the result would be identification of teacher needs, collaboration on professional development plans to meet those needs, and delivery of truly systemic professional development.

A second step toward improving professional development would be to deal with

the reality that "there ain't no Santa Claus!" Inservice teacher education is no different than preservice teacher education in that somebody has to pay. In preservice education, the sources of funds are tuition, state allocations (usually based on student Full Time Equivalent [FTE] enrollment), and federal funds. In most cases, state allocations are the largest source of funds, followed by tuition, and finally federal funds. Both higher education and public schools receive little state funding for inservice education and usually only Eisenhower funds as a federal source. Local funds for inservice education are usually quite limited or non-existent.

No doubt there is a need for more money to provide quality professional development activities for our science teachers, but given the existing political climate, the reality is that those additional funds are not likely to forthcoming. An alternative is to employ some creative ways to use the resources that are available to us. A good start might be to discard the myth that teachers have to depend on someone else for professional development. The *National Science Education Standards* (National Research Council, 1996) states, "The traditional distinctions between 'targets,' 'sources,' and 'supporters' of teacher development activities are artificial." All good teachers provide a substantial portion of their own professional development through reading (Clough, 1993) and other means of independent study. Granted, this may not be the best way for some teachers, nor should it be the only avenue available. But for many of our best teachers it provides a way to be on the cutting edge and to do so at their own convenience. These bold explorers can and should become resources for other teachers who prefer more directed forms of professional development.

Even through independent study might be the primary mechanism for only some teachers, it is a mechanism that should be available to all teachers. To make that happen, professional development planners must ensure that books, professional journals, video tapes and other independent study resources are available. One way of doing this is through institutional memberships in professional organizations in which the journals come directly to the schools. This is an especially effective mechanism in elementary schools where teachers are usually responsible for more than one subject area. Schools should provide some funds for these memberships, but teachers should contribute to them as well. This approach provides the fruits of many professional memberships at a fraction of the cost.

Video tapes are attractive professional development alternatives for many teachers. In a larger school system, tapes can be purchased based on needs surveys of the teachers and used by individual teachers or groups of teachers within the system. For smaller systems a collaborating video tape library might be formed to serve a number of such systems.

Teachers who have become expert in an area or who have developed a novel lesson, activity plan, unit, or procedure should use a camcorder to record the explanation and demonstration of their innovation and add these tapes to the professional development library. We tend to think that video tapes to be used for professional development must be elaborate productions costing thousands of dollars. In fact a great deal of professional development could be delivered through homemade video tapes that are done with a home or school camcorder and simply have good lighting, clear speech, and a comprehensive explanation/demonstration of the procedure or innovation.

As computer networking becomes a more integral part of the educational world there is the opportunity to take advantage of this technology as a professional development tool. (Monty, 1994) Widely touted as a powerful classroom tool, the Internet has yet to live up to that billing; however, it is an extremely valuable teacher resource and can play a vital role in professional development activities. An especially good example of the potential of this technology is the World Wide Web. Multimedia browsers for the WWW, such as Netscape and Mosaic, have given educators a nearly intuitive interface to a myriad of resources. Numerous government and professional organizations such as the National Science Foundation, the National Science Teachers Association, most major universities, and many public schools maintain Web sites that often offer a wide variety of information and resources.

In an ideal world, each teacher would have access to a high-quality network and would use it daily. Those uses would include using e-mail and discussion groups to communicate with colleagues; searching databases to find lesson plans, funding opportunities or other resources; or checking calendars for training opportunities. Unfortunately, a lot of schools do not have access to the Internet. But for those that do, it would take no more than one science teacher to dig out those valuable resources and share them with others.

But the traditional consumers of professional development activities can't be expected to carry the entire improvement burden. Historically, the most common primary providers of professional development have been university faculty going into the public schools, or in many cases public school faculty attending the university for professional development activities. More often than not, two sets of circumstances have brought university faculty into the public schools, or public school faculty to the university. The first is the "get a grant/hold a workshop" scenario. The second set of circumstances is the one-shot, short-duration workshop conducted by the university expert at the request of the school system. The grant-workshop model has been supported for years by government agencies and private foundations, and although some very fine things have been accomplished through these programs, they have not nor will they ever accomplish either goal four or five. The same can be said for the one-shot workshops. Most of these programs have had limited and short-term impact on the systems from which the participates originate because the programs were not part of a more systemic, and sustained effort.

Universities are not entirely to blame. This systemic and sustained effort has not historically existed at the national and state levels, even though the National Science Foundation's recent State Systemic Initiatives represent attempts to do just that. However, long-standing turf protection battles between and among universities have often created a stand-off when collaboration is the desired result. Perhaps university faculty and administrators need to learn to work in a productive consortia in which a number of universities collaborate with public schools state and federal agencies, and private entities to plan and deliver professional development activities that are systemic and sustained.

If significant progress could be made in improving professional development opportunities for science educators, what characteristics might the resultant activities possess?

• Every science teacher would be involved in professional development activities each year.

• Professional development goals and activities would be directed toward meeting well documented needs.

• Professional development activities would be funded at a level sufficient to support high quality programs.

• Greater use would be made of the abilities of innovative teachers to provide professional development for their peers.

• Every university would have professional development partnerships with public schools in their service area, and higher education personnel should be participants in as well as delivers of professional development activities.

• Extensive use would be made of technology to share professional development resources.

• Programs would be conducted that are especially designed for the new science teacher as well as their more experienced peers.

• State, regional, and national conventions of professional science education organizations would be an integral part of professional development plans.

• The bulk of professional development would be delivered on professional time, not personal time.

• Special attention would be given to the needs of the rural, often isolated, school systems.

The time has come to look critically at methods that have become an integral part of professional development activities and to judge them in terms of how well they have worked. Professional development activities are presently addressing a rather large number of science education needs that were being addressed 30 years ago, and the measures being used now are much the same as those that have failed to produce the desired results over that 30 year span. Both preservice and inservice education have failed to produce science education experiences for our students that are universally hands-on, inquiry oriented, and student centered, and that stimulate students to want to learn more science. Staff development activities aren't the only reasons science teaching hasn't undergone greater change, but perhaps some radical changes in the approaches to staff development are worth a try.

## References

American Association of State Colleges and Universities, (1988). Junior High Middle School Science Improvement Project, Washington, DC.

Clough, Michael P. (1992, October). Research is Required Reading. *The Science Teacher*, 59(7), 36–39.

Goals 2000: Educate America Act (1994), Pub. L. No. 103–227 (3/31/94), State. 108.

Monty, Ken (1994). Staying on Top: Professional Development for Educators. *The ELLIPSE*, *3(2)*, 1.

National Research Council. (1996). *National Science Education Standards*. Washington, DC: National Academy Press.

## Author Note

**Don H. Kellogg** is director of the Center of Excellence for Science and Mathematics Education and professor of educational studies at the University of Tennessee at Martin. He has written and directed science, technology, and environmental education grants and designed inservice programs for over 27 years.

# Developments in Laboratory Safety

Jack A. Gerlovich

This paper is an update of Downs, Gerard, and Gerlovich's 1988 monograph entitled, "School Science and Liability." In the intervening eight years there has been a proliferation of federal and state legislation, as well as position statements by professional science education organizations, impacting all facets of safety in science settings. These changes focus in three major domains: legal liabilities, safety management, and chemical management. This paper's purpose is to assess science educators' understanding of these elements of science safety and to provide updates regarding them.

In a study by Gerlovich (1995), 300 science educators (teachers, supervisors, and college science education professors) were surveyed regarding science safety issues in the following ten states: California, Illinois, Louisiana, Mississippi, New Jersey, Oklahoma, Pennsylvania, Tennessee, Texas, and Utah. The surveys were conducted as part of science safety inservice training programs being conducted by the author. Table 1 provides a listing of the content assessed and numbers of correct responses by participants.

## Legal Liabilities

In the area of legal liabilities, Gerlovich discovered that science educators have a very poor understanding of liabilities and professional responsibilities. In the area of legal liability, only 45 percent (136 of 300) of the participants knew anything about *tort law* and its focus upon personal physical injury cases. Exactly 9 percent (27 of 300) of participants knew *sovereign immunity*, within tort law, was once used as a legal defense in many states. Only 8 percent (25 of 300) of the participants knew that the *save harmless provision* is currently used as a powerful defense for educators in tort negligence cases. Only 22 percent (66 of 300) of participants could explain what constituted *negligence* in science education. Only 6 percent (19 of 300) of participants knew that *due care* was a synonym for assuring that the educator was not negligent, and that it consisted of satisfying three major duties.

## Safety Management

In the area of safety management, only 16 percent (48 of 300) of participants assessed knew the function of *ground fault interrupters* (GFIs). Fourteen percent (41 of 300) of participants knew the best type of fire extinguisher to place in science labs. Exactly 4 percent (12 of 300) of participants knew the American National Standards Institute (ANSI) Z87 marking on approved eye safety equipment. Only 15 percent (58

**Table 1.** *1994 Science Safety Knowledge Survey of 300 Science Educators from 10 States.*

| ITEM | CORRECT RESPONSES |
|---|---|
| **Legal Issues** | |
| Tort law | 136 |
| Sovereign immunity | 27 |
| Save harmless provision | 25 |
| Negligence | 66 |
| Due care teaching duties | 19 |
| **Safety Management Issues** | |
| Equipment | |
| Ground fault interrupters | 48 |
| Fire extinguishers | 41 |
| Eye protective equipment | 12 |
| Placement of strategic safety equipment items | 58 |
| Facilities | |
| Entrances & exits | 212 |
| Procedures | |
| First aid for chemical splash to eyes | 63 |
| Class size limitations | 55 |
| **Chemical Management Issues** | |
| Storage | 74 |
| Right-to-Know legislation | 3 |
| Chemical Hygiene Plan legislation | 4 |
| MSDS | 40 |

**Note.** From, Gerlovich, J.A. (1995). *Was I Supposed to Know That? Science Educators understanding of Science,* Vol. 6, 3. Sept. 1995.

of 300) of participants knew that, unless other circumstances dictate, chemical exhaust hoods should be placed as far away from primary lab entrances/exits as possible. Nearly 71 percent (212 of 300) of participants knew that multiple lab exits are essential for science safety. Only 21 percent (63 of 300) of participants knew that, in the event of a chemical splash to the eye, medi-cal experts recommend flushing with temperate, aerated water for 15 minutes. Only 18 percent (55 of 300) of participants knew the recommendation of the National Science Teachers Association (NSTA), and now supported by numerous other science education professional science organizations, for class/lab size limitations.

## Chemical Management

In the area of chemical management, 25 percent (74 of 300) of participants knew how chemicals should be stored for safety. Only 1 percent (3 of 300) of participants could identify the major requirements of the Occupational Safety and Health Association (OSHA) Hazard Communication Standard or *Right-to Know* (RTK) legislation, pertaining to hazardous chemicals in the workplace. Only 1 percent (4 of 300) of participants could identify the major components of the OSHA Chemical Hygiene Plan. Only 13 percent (40 of 300) of participants knew the purpose of Material Safety Data Sheets (MSDS).

These statistics are indeed discouraging when one realizes that science educators can avoid most negligence allegations resulting from accidents if they (a) anticipate safety problems and attempt to have them corrected in the most expeditious manner practicable; (b) conform to applicable laws (Right-to-Know, Chemical Hygiene, goggle legislation; codes—plumbing, electrical, fire, architectural) and professional standards (NSTA instruction, supervision); and (c) perform their science education duties.

## ANTICIPATE SAFETY CONCERNS

A simple, effective safety philosophy developed by Gerlovich (1994) states that teachers should weigh all activities involving students for their educational value versus potential hazards. If the activity is designated educationally valuable, but potentially dangerous, the teacher has three options: (a) add more safety equipment and/or precautions; (b) limit the activity to a teacher demonstration; or (c) eliminate the activity entirely. At the elementary level, for instance, what should be a teacher's decision regarding the inclusion of the potassium dichromate volcano activity? At the secondary level, what about blood typing activities?

One effective way to assure that safety issues are addressed consistently and effectively is to develop, implement, and enforce science department policies (Gerlovich, Hartman, and Gerard, 1994) that reflect input from all staff members. This is sometimes best accomplished by performing a thorough safety assessment of techniques, equipment, and facilities, looking for consistent concerns. These would then be rank ordered and compared for patterns. From these, department policies would be developed. Sample policies should include, but in no way be limited to:

*It is the policy of the_____ science department to:*

1. conform to all applicable laws, codes and standards of this state.

2. conform to all applicable standards of the science education profession.

3. use science rooms only for their designed science functions.

4. apprise the administration, in writing, of all potential and actual hazards.

5. provide students with instruction appropriate for the science curriculum.

6. provide students with appropriate supervision for all science activities.

7. maintain the science teaching and learning environment in a safe condition.

These universally supported, general statements must then be placed in writing,

endorsed by the entire department, and communicated to the school and/or district administration. In nearly all cases, such statements are well received by the school and/or district administration. Specific written requests for corrections of safety concerns would then follow to insure that the policies are met. These statements must be specific and matter-of-fact. For example, "Room 136 has 36 students enrolled in third period chemistry; this is an unsafe situation and violates our district science safety policy regarding supervision and conformity with science education professional standards."

## CONFORM TO LAWS, CODES, AND STANDARDS

Safety should be everyone's concern. People who behave in an unsafe manner not only expose themselves to a greater risk of accident or ill health, but increase their chance of being held legally liable for injury to the property and health of others. For example, a careless chemistry teacher endangers more than his or her own life; by also endangering the lives of others, he or she increases the chances of being legally liable in an accident. There is a direct correlation between safety and the conduct the law and our profession expects of us.

The civil law expressed throughout the 50 states presumes that people have an obligation to behave reasonably where their conduct affects other people. This is an extension of the criminal law, which is designed to protect the lives and property of citizens. The proscriptions embodied in criminal law are considered undisputed specific dictates of reason. But in a free society, it is neither possible nor desirable to specify everything that citizens ought to do to preserve order within society. Therefore, civil law evolved to further protect the lives and property of individuals from events that are not criminal, but may nonetheless be unreasonable. The protection is afforded through a process of awards to victims of unreasonable behavior, exacted in the form of judgments against those who caused damage by behaving unreasonably or unsafely. The *save harmless provision* (State of Iowa, 1992) is nearly universally applied in all states. It generally states that accidents can happen, parties can be injured by educators, and law suits can be filed against the educator. However, unless it can be proven that the educator broke the law (goggle legislation, Right-to-know legislation, etc.), or was grossly negligent (violated well accepted professional organizational guidelines, established codes, department of education standards, etc.), he or she would be defended to the limits of the resources of the school district or state.

Governmental bodies, such as school districts, are sometimes protected from being held liable by a doctrine called *sovereign immunity* (Joyce, 1978). In the jurisdictions that follow this doctrine, tort negligence cannot be charged to the governmental employer, but can to the employee who is negligent. Many states have abandoned this doctrine in favor of allowing tort suits against governmental bodies. Most jurisdictions provide that the governmental agency must defend and indemnify, or compensate, the employee for such suits in any case. This is known as a hold or *save harmless provision*. However, such provisions do not apply in cases where the employee is accused of a willful tort, i.e., breaking the law or behaving in a grossly negligent manner.

The legal principles that govern this civil form of accountability are much the same throughout the country and apply uniformly to all sorts of conduct (Rice, Strope, Brown, 1981). The most basic principle is that one has a duty as a citizen to behave reasonably toward others, which translates into legal terminology as a duty owed to avoid negligent behavior. The law defines *negligence* as acting differently than a reasonable person would in a specific circumstance. It can mean either doing something unreasonable (commission) or failing to do something reasonable (omission). It has often been defined as "conduct falling below a standard set by the law to protect others from harm, or lack of due care," (*School Laws of Iowa*, 1992). Due care, in turn has been subdivided into three teacher duties, which will be explained later in this paper. It may be helpful to think of negligence as carelessness. Simply put, anyone can be held legally accountable for the consequences of his or her unsafe actions. For those in science education, the lesson is that unsafe conduct in the laboratory may make one personally and financially responsible for any harm resulting from such conduct.

The standard of accountability in any given circumstance is determined by what ordinary people (often a jury) deem to be reasonable. There usually exists a body of law, built up over time from previous similar cases, that gives precedent to the present case. In arriving at its decision, a jury is guided by the opinions of experts familiar with the specific circumstances in which the accident took place. These experts testify as to whether certain conduct was reasonable in a particular situation. For example, if a lawsuit arises from an accident occurring in a chemistry laboratory, expert witnesses such as leading chemistry teachers and safety experts may testify as to whether the accident was caused by unreasonable conduct, and if so, by whom. A body of case law then builds up around specific sets of facts, guiding the decisions in future cases with similar sets of facts (Gerlovich, Gerard, 1989).

How then should science educators conduct themselves to avoid being found negligent and held liable for damages arising out of an accident? Although the law does not say that accidents always result from negligence (they often happen through the fault of no one), there is a tendency in a complex technological society to try to assign fault and apportion liability accordingly. This is especially true in education, where students are presumed to be under the care and protection of their teachers. A science educator, then, must conform to what other science educators would consider reasonable in order to minimize his or her risk of legal liability. This recommendation is largely common sense. For example, a science teacher who asked a student to pour water into concentrated acid to demonstrate its explosive effect would surely be found negligent. More technical matters, however, require more judgment and testimony by other science teachers who are familiar with the given situation. The operative questions revolve around what the consensus of opinion would be as to the reasonableness of this conduct by others in the science education profession. For this reason it is of the utmost importance that educators keep abreast of the developments in their profession in terms of safety. Continuing education is essential to staying within the mainstream of opinion as to what constitutes reasonable behavior.

## PERFORM YOUR DUTIES

Science educators must accept their responsibility to implement safe conduct by satisfying three professional and legal duties. First is the duty to adequately supervise students —to keep a watchful eye. Improper supervision is the single greatest cause of legal liability in science education. Proper supervision will prevent most accidents. The second duty is to instruct students relative to all foreseeable and reasonable hazards. A teacher must tell students all that they need to know to participate safely in any given learning project. The appropriate and accurate dissemination of information relevant to student safety should be an integral part of lesson planning. The third duty is to maintain safe equipment and a safe environment in which students learn. All three duties pertain to the teacher's responsibility when in charge of students in a learning situation. Again, the duties must be performed accord-

ing to the acceptable standards of the profession to insure protection from legal liability.

Consistent with adopted policies, the science supervisor and teacher should take active roles in planning and conducting safe science activities. The amount of supervision a teacher provides will vary with the degree of danger involved in the activity, the novelty of the environment to be experienced, and the ages and special needs of the students participating. As a professional guideline, the NSTA (1991) recommends a teacher-student ratio of 1:24 in classroom and laboratory situations and 1:10 during field experiences. Signed permission forms should be received from parent(s) or guardian(s) for each student before his or her involvement in field experiences. On-site visits of field study areas should be conducted by the teacher and supervisor, noting any potential and actual hazards and communicating these to students within the context of their work. It is also wise to speak to the landowner concerning any subtle hazards. Appropriate school-sanctioned student transportation must be provided, as well as insurance. The administration should be informed of all off-school experiences in order to reduce liability. Students should be paired into buddy system teams to help protect each other. Some states even have statutes regarding school field trips. According to Troy and Schwaab (1981) the general features of such laws included provisions for transportation, liability insurance, proper supervision, and attendance.

Science teachers and supervisors must see that students receive appropriate safety instruction for their activities. This is critical for hands-on laboratory activities. Student safety contracts have proven to be effective tools for apprising students of responsibilities and safety components. Such contracts should include orientation to safety equipment and safety procedures, including evacuation of the environment and proper use of safety equipment items. Monitored simulations of foreseeable emergencies with students have also proven to be effective learning techniques. Knowing who is wearing contact lenses can also be valuable in an eye emergency. Students should be provided an overall orientation to the contract on the first day of class. However, each component should be emphasized as necessary within the structure of the course. Student understanding of each component should be verified by his or her signature and date at that time. Parent(s) and/or guardian(s) should also sign the contracts indicating their support and recognizing the need for reasonable rules.

Safety signs are another effective communication technique. Signs placed in conspicuous places within labs and classrooms help to keep the safety message before students. Examples are signs requiring the use of safety goggles; the location of fire extinguishers, fire blankets, eyewash, and drench showers; and master shutoffs for water, gas, and electricity.

It is also the responsibility of the teacher and science supervisor to properly maintain the teaching and learning environment. It is vital that teachers and supervisors conduct regular assessments of safety techniques, equipment items, and instructional activities (Gerlovich, 1994, Yohe, 1992). Any identified potential hazards should be written up and passed to the administration as part of the department policy to provide a safe teaching and learning environment. It is imperative that proper and appropriately-sized safety equipment items be provided to teachers and students and that they know how to wear, use, and clean such items. Among the more notable general science safety items are functioning fume hoods, fire blankets, fire extinguishers, eyewashes, drench showers, GFI protected electrical outlets, and emergency telephone numbers. Chemicals must be properly labeled, stored, and disposed of according to federal, state, and local guidelines. Teachers and supervisors must understand

and comply with Eye Protective Equipment, Right-to-Know, Laboratory Standard, Chemical Hygiene, and Bloodborne Pathogen requirements as applicable for their state.

Statutory mandates and/or guidelines (State of Iowa, 1992) issued by departments of education regarding eye protective equipment generally state that ANSI approved equipment must be worn whenever the potential for an eye injury exists. Z87 and/or manufacturers' trademarks placed on lenses and/or molding indicate compliance with this standard. It guarantees that the equipment will not burn or break under normal use. Science educators should err on the side of caution when interpreting circumstances requiring the wearing of such protective equipment. They should act as role models for students by wearing the eye protective equipment themselves. They should insist that the equipment is kept clean and defect free.

Increasing numbers of organizations are now encouraging the wearing of contact lenses in science laboratories *If* they are worn under ANSI approved goggles which seal to the face and are unvented. The OSHA (1994) Personal Protective Equipment (PPE) for General Industry Standard provides the following statement regarding contact lenses as part of its preamble:

OSHA believes that contact lenses do not pose additional hazards to the wearer, and has determined that additional regulation addressing the use of contact lenses is unnecessary. The Agency wants to make it clear, however, that contact lenses are not eye protective devices. If eye hazards are present, appropriate eye protection must be worn instead of, or in conjunction with, contact lenses.

As of 1995, the American Chemical Society had no official policy statement on the use of contact lenses in the workplace where chemicals are used. However, in *Safety in Academic Chemical Laboratories*, they do make the following statement:

> Wearing of contact lenses in the laboratory is normally forbidden because contact lenses can hold foreign materials against the cornea. Furthermore, they may be difficult to remove in the case of a splash. Soft contact lenses present a particular hazard because they can absorb and retain chemical vapors. If the use of contact lenses is required for therapeutic reasons, fitted goggles must also be worn.

Even the Canada Safety Council has stated (1985) that is has learned of no adequately documented proof that the use of contact lenses on the job presents any hazard.

The OSHA Right to Know legislation (Thompson, 1990, OSHA, 1994) became effective in 1985, supported by the ANSI, National Fire Protection Association (NFPA), and the Department of Transportation (DOT). It is targeted at all private schools and selectively for public schools by state. The legislation requires that (a) teachers and supervisors know chemical hazards (Material Safety Data Sheets—MSDS provide this information from manufacturers); (b) written plans for handling hazardous chemicals are available and implemented; (c) chemicals are labeled and properly stored (by chemical family, protected against heat extremes, properly ventilated, secured from unauthorized access); (d) the public, and emergency services personnel, are provided information on these chemicals as requested; and (e) employees receive training regarding their local plan. Teachers and supervisors, who are in the process of complying with this or Chemical Hygiene legislation may want to investigate the Canadian Centre for Occupational Health and Safety's (CCOHS, 1994) collection of 100,000 MSDS, via the Internet.

OSHA's Laboratory Standard legislation became effective in 1991 and requires the development and implementation of written chemical hygiene plans (Hall, 1994) that (a) set forth procedures, equipment, personal protective equipment, and work practices that protect lab workers from hazardous chemicals used in that lab; (b) are made available to employees, their designated representative, and the Assistant Secretary of Labor upon request; (c) are reviewed and updated at least annually; and (d) should address the specifics of standard operating procedures, chemical hazard control measures, laboratory ventilation, employee information and training, prior approval for lab use, medical consultation and examination, designation of a chemical hygiene officer, and any other special procedures necessary.

The laboratory standard does not apply to uses of hazardous chemicals that do not meet the definition of laboratory use, and in such cases, the employer shall comply with the relevant OSHA standard in 29 CFR 1910, even if such use occurs in a laboratory. As of January 31, 1991, laboratories engaged in activities that are encompassed within the definition of laboratory use must have in place a written Chemical Hygiene Plan (CHP) outlining how the facility will comply. This OSHA standard applies to all employers engaged in the laboratory use of chemicals. Laboratory use means "chemicals are manipulated on a laboratory scale where the chemicals are handled in containers designed to be safely and easily manipulated by one person; multiple chemical procedures are used; procedures are not of a production process; and, protective laboratory equipment and practices are in common use to minimize employee exposure."

OSHA's Bloodborne Pathogen (OSHA, 1992) legislation became effective in 1991. The purpose of the legislation is to (a) minimize employee exposure to Human Immunodeficiency Virus (HIV) and Hepatitis B through contact with the bodily fluids of others; (b) provide prophylactic equipment items (masks, rubber gloves, etc.) to employees as protection from exposure to human bodily fluids; and (c) see that employees with a higher potential for exposure to others bodily fluids, due to the nature of their work, receive inoculations for Hepatitis B and other communicable diseases. Science educators must be cognizant of potential exposure of students, peers, and themselves, and help guard against it. In many instances, lesson plans involving students directly will need to be adjusted or eliminated. Supervisors must keep subordinates and peers apprised of changes in the legislation as well as safety adjustments within the science curriculum.

As part of department policies, teachers and supervisors should assure that all applicable fire, electrical, and architectural codes (Ashbrook, Renfrew, 1991, Saunders, 1993, DiBerardinis, 1993) are met. All fire codes for appropriate fire extinguisher type (ABC triclass or halon), size, and location; prevention of overcrowding; and evacuation routines, should be complied with. Electrical codes must be satisfied, addressing, at a minimum, the location and proper use of master electrical shut-off switches, GFIs installed on strategic outlets, and the placement of electrical outlets so that extension cords are not required. Plumbing codes should assure the proper location of student gas jets as well as convenient location of master natural gas shut-offs. In addition, such codes should address chemical exhaust hood material composition (chemically inert), blower speed and type, hood placement (generally as far away from primary room entrances as possible), air intake and exchange of fresh air, and exhaust gas effluent. Architectural code compliance should include laboratory and storeroom design, floor space, occupancy limitations, placement, size and number of exits, minimal ventilation, smoke and fire protection equipment, and appropriate-

ness of chemicals being stored and used in the facilities.

In addition, nearly every state has implemented rule-specific safety requirements, such as those regarding eye protection. Failure to comply with these laws may result in an automatic finding of negligence against the teacher who does not comply. In many states it is the responsibility of the teacher to remove any student who persistently ignores the rules to wear such specific safety equipment as goggles. This should be a conspicuous part of the student safety contract. Teachers and supervisors must assure that goggles meet ANSI standards for burn and breakthrough. Purchasing agents must be apprised that only goggles meeting the above ANSI Z87 standards will be accepted. It is also imperative that educators remain current regarding subtleties of eye safety such as contact lenses. The current consensus (Preuss, 1995, Segal, 1995, Cullen, 1995) is that contact lenses can be worn during science activities if adequate goggle protection (unvented, antistatic, antifogging lenses meeting ANSI standards) is worn over them. Student safety contracts should include a component regarding the wearing of goggles and contact lenses.

## SUMMARY

Science supervisors and teachers are no different from other professionals when it comes to avoiding legal liability while assuring quality in their educational curriculum. They must conform to all applicable laws and codes, behave reasonably, use good common sense, remain current in their field, keep the administration informed of hazards and needed change, employ safe practices recommended within the profession, and avoid exposing students to hazards deemed unacceptable by qualified peers.

## References

American Chemical Society (1990). *Safety in Academic Chemistry Laboratories*, 5th ed. American Chemical Society, Washington, D.C.

Asbrook, P.C. & Renfrew, M.M. (1991). *Safe Laboratories: Principles and Practices for Design and Remodeling.* Chelsea, MI: Lewis Publishers, Inc.

Cullen, A.P. (1995). Contact Lenses Emergencies. *Chemical Health & Safety*, 2, 1, 22–25.

Canadian Center for Occupational Health and Safety (CCOHS). *Chemical Database* (on-line service, call for information). Ontario, Canada: Author.

DiBerardinis, et al (1993). *Guidelines for Laboratory Design*, 2nd ed. New York, NY: John Wiley & Sons, Inc.

Downs, G.E., Gerard, T.F., Gerlovich, J.A. (1988). School Science and Liability. In *Third Sourcebook for Science Supervisors*, by Motz, L. M. & Madrazo, G., Jr., pp121–126. Washington, D.C.: National Science Supervisors Association.

Gerlovich, J.A. (1995). *Was I supposed to know that? Science educators understanding of science safety issues.* Science Education International, 6,3. Sept. 1995.

Gerlovich, J.A. (1993). Some safety concerns in school science settings and their implications for science supervisors. *Science Education*, 2, 1, 28–31.

Gerlovich, J.A., Hartman, K., Gerard, T. (1994). The Total Science Safety System, 7th Edition [Computer Software]. Waukee, IA: Jakel, Inc.

Gerlovich, J., Gerard, T. (1989). Keep Science Experiments Safe and Students Sound. *American School Board Journal*, 176, 5, 40–41.

Hall, S.K. (1994). *Chemical Safety in the Laboratory.* Boca Raton, FL: CRC Press, Inc.

Joyce, E.M. (1978). Law and the laboratory. *The Science Teacher*, 45,6, 23–25.

National Science Teachers Association (1991). *Guidelines for Self-Assessment.* Washington, D.C.: Author.

Occupational Safety and Health Administration (1994). OSHA Regulations, documents, and technical information (CD-ROM). Washington, D.C: U.S. Department of Labor.

Occupational Safety and Health Administration (1991, January 31). Occupational exposures to hazardous chemicals in laboratories, final rule (CFR, Part 1910). *Federal Register.* Washington, D.C.: U.S. Government Printing Office.

Preuss, A. (1995). The World of Contact Lenses. *Chemical Health & Safety*, 2, 1, 12–15.

Rice, D.R., Strope, J.L., Brown, C.L. (1981). Test your legal liability. *The Science Teacher*, 48, 5, 44–45.

Saunders, G. Thomas (1993). *Laboratory fume hoods: A user's guide.* New York, NY: John Wiley & Sons, Inc.

Segal, E.B. (1995). Contact lenses and chemicals: an update. *Chemical Health & Safety*, 2, 1, 16–21.

State of Iowa (1992). *School Laws of Iowa.* Des Moines, Iowa: Iowa Department of Education.

Stones, I. & Spencer, G. (1985). *Contact lenses in the workplace: the pros and cons.* Ontario, Canada: Canadian Centre for Occupational Health and Safety.

Thompson, G.R. (1991). *Compact school and college administrator's guide for compliance with federal and state right-to-know regulations.* Philadelphia, PA: The Forum for Scientific Excellence.

Troy, T.D. & Schwaab, K.E. (1981). Field trips and the law. *School Science and Mathematics*, LXXXI, 8, 689–692.

U.S. Department of Transportation (1994). *Code of Federal Regulations* (CFR), 49-Transportation, Parts 100-177. Washington, D.C.: U.S. Government Printing Office.

U.S. Department of Labor, Occupational Safety and Health Administration (1992). *Code of federal regulations* (CFR) 29, Part 1910.1030. Occupational Exposure to Bloodborne Pathogens, Subpart Z Bloodborne Pathogens Standard Summary Applicable to Schools. Washington, D.C.: U.S. Government Printing Office.

Yohe, B. & Dunkleberger, G.E. (1992). Laboratory safety and inspection procedures. *Journal of Chemical Education*, 69,2, 147–149. Washington, D.C.: American Chemical Society.

## Author Note

Jack A. Gerlovich is an associate professor of science education at Drake University. He has written and directed science and technology grants worth over $5 million. He has taught science at all levels of education for 25 years. He has also published widely in science safety, including two comprehensive software packages.

# Enhancing the Value of Business and Education Partnerships: Setting and Meeting Higher Science Education Standards

Lawrence B. Flick
Norman G. Lederman
Elizabeth A. Lambert

Business and education have pro-claimed themselves to be *de facto* partners for most of this century. A major tenant of this partnership was established when the *Cardinal Principles* (Bureau of Education, 1918) defined American high schools as prototypes of democracy whose goal was to prepare students to be good citizens. Being a good citizen was linked in part to procuring a fulfilling vocation. To that end, they recommended that schools were to employ those successful in vocations as instructors and the "actual conditions of the vocation should be utilized either within the high school or in cooperation" (p. 7) with business and industry. At this pivotal time, science and math education were also being examined and schools were charged to make instruction more practical with more direct contact with practical problems and concrete materials (Committee of Ten, 1896).

Teachers have traditionally invited the occasional guest speaker or taken field trips where inclination and budgets allowed. However, for both fiscal and curricular reasons, schools and businesses should become serious about the form and function of their inevitable partnership. The classroom is in part a way to cloister students away from societal distractions while developing skills and examining a broad range of ideas. But the nature of our technological and information-rich society demands that schools help students assume the role of citizens with marketable skills (National Commission on Excellence, 1983). High paying, low skilled jobs are disappearing and competencies necessary in the workplace have direct implications for schools (U.S. Dept. of Labor, 1991). Goal five in the National Educational Goals Report (National Educational Goals Panel, 1991) states that "By the year 2000, every adult American ... will possess the knowledge and skills necessary to compete in a global economy ..." (p. 237). One of the objectives for Goal five is "Every major American business will be involved in strengthening the connection between education and work" (p. 237).

Setting higher and better articulated standards in science education will require that students and teachers have regular access to the world beyond the classroom. "Relationships should be developed with local businesses and industry to allow students and teachers access to people and institutions, and students must be given access to scientists and other professionals ... to gain access to their expertise and the laboratory settings in which they work" (National Research Council, 1996, p. 221). Educational reform, especially in math and science, has virtually mandated that technical experts engage in teaching activities. However, there has been minimal program evaluation concerning the educational effectiveness of partnerships. Yet, programs for developing these teaching relationships, where students are prepared to work with experts, where teachers have integrated the experiences into the curriculum, and where outside experts have learned enough about the complexities of teaching, are just beginning to take shape.

## A TYPOLOGY OF PROGRAMS

Many schools and businesses have been actively working to create partnership models. Partnership has a variety of meanings, but key elements of most programs include identification of mutual goals, an organization whose leader is a spokesperson for the partnership, and a shared understanding for how the activities of the partnership are to be funded. A review of national responses that involved education and community partnerships indicated that initiatives can be taken at many levels, whether supported by modest resources or millions of dollars. Reviews of alliances, collaboratives, and partnerships have been made for both large and small programs (Atkin & Atkin, 1989; Maeroff, 1988). What follows is a typology of programs derived from this review and from the authors' experiences.

### Resources for Teachers

Businesses can offer opportunities for teachers to improve their knowledge or skills in a particular area. During summers or at other release times for teachers, community expertise can be shared with teachers in seminars or workshops directed at particular topics or training sessions. These sessions could be concerned with developing skills with equipment or techniques, examining technical or managerial systems, or under-

standing the tasks and responsibilities of the company.

### Resources for Teaching

Related to the previous item, business expertise and facilities might be made available as a resource helping teachers develop teaching ideas more directly linked to workplace examples. Some teachers develop personal networks for seeking out information when needed. More formalized relationships could be developed to facilitate inquiries and share expertise, such as disseminating electronic mail addresses or creating local lists or bulletin boards.

### Adjunct to the Classroom

Perhaps the most common form of using community resources is by inviting guest speakers into the classroom or taking students on site visits. Programs such as speakers bureaus and community resource directories have institutionalized this type of activity (Flick, Fekete, Hawkins, & Stone, 1995). This direct interaction with the classroom should be strengthened through a more systematic sharing of educational purposes and goals with community resource people. This will be discussed more fully in the section Planning for Success.

### Career Education

Business could offer paid or unpaid intern positions so that teachers can learn more about technical careers. Similar arrangements can be made for students as part of their school coursework. Reform documents in science and mathematics education emphasize the importance of presenting the nature of the scientific enterprise as well as increasing awareness of careers in technical fields (American Association for the Advancement of Science, 1993; National Research Council, 1996; National Educational Goals Panel, 1991). Direct contact in the form of on-the-job observation or experience for teachers and students creates a richer understanding of the way skills, knowledge, and attitudes are integrated in actual work settings.

"Shadowing" programs have created opportunities for teachers to follow scientists, engineers, and technicians in the workplace as well as allowing workers in technical fields to shadow teachers. These programs have helped people in the technical and scientific workplaces understand the constraints and rewards of teaching. It has also increased teacher networking with community resources.

## Teacher Education

A business can work with a teacher education institution in order to collaboratively develop projects for interacting with schools. For example, Lawrence Livermore National Laboratories (LLNL) and Oregon State University (OSU) are developing a partnership that will potentially address several interrelated goals that reach beyond traditional arrangements. This program will formalize a collaboration between the Science and Mathematics Education Department at OSU with the Education Program at LLNL to address activities in four areas: (a) teacher development and curriculum development, (b) graduate student research, (c) teacher research internships, and (d) outreach and informal science education. The outcomes will include improved science education programs in schools and improved research and evaluation opportunities in teacher education.

## Business and Education Compacts or Consortiums

Community, business, and education leaders have formed organized groups called compacts or consortiums that meet regularly to initiate or facilitate the above activities. These groups are often formed as a result of identifying a common need, for example to improve science and math education to meet state education goals. These groups are also created on the heels of a successful program where the participants become aware

that an expanded partnership can increase the number of successful programs.

Elements of successful collaborations include a clear and very specific mission statement and concrete goals. The Business Education Compact (BEC) based in Beaverton, Oregon, has been serving the Portland metropolitan area for ten years. Their mission is to "walk the talk of education reform by setting up successful grass-roots partnerships between business and education leaders who are committed to excellence in every classroom" (Northwest Regional Educational Laboratory, 1994/95, p. 3). Their goals are to:

• Foster programs that promote student excellence in math and science education and careers in science, engineering, and the technical fields, particularly for young women and minorities.

• Create better connections between the classroom and the workplace.

• Reinforce the concept that learning spans a lifetime by reducing barriers to education and by providing incentives for all citizens.

It has been the experience of the BEC that businesses rarely hesitate to establish partnerships with schools, but more school administrators need to support teacher participation. The BEC model has been used to start similar compacts in Oregon and throughout the Northwest. An example of a specific project that helped catalyze a partnership involved the development and evaluation of an expanded directory of community resources for use by teachers in south central Washington state (Flick et al., 1995).

## PLANNING FOR SUCCESS

The effectiveness of partnerships on schools can be examined in three ways: (a) effects on teachers and their teaching strategies, (b) effects on what students learn, and (c) effects on schools or districts (Kubota,

**Table 1.** *Outline of Program for Enhancing Interactions among Community Resource People, Teachers, and Students*

I. Prepare a handbook of successful practices and video-taped examples for use by community resource people and teachers.

a. Discuss the handbook with teachers and community resource people to identify important issues.

b. Solicit sample materials from successful presentations and site visits.

c. Design generic forms and checklists that facilitate communication about educational goals and their relationship to community expertise.

d. Produce the handbook in a loose-leaf format so that users can personalize its contents.

e. Video-tape successful presentations that emphasize interactive discussions and hands-on activities. Edit with input from teachers, administrators, and community resource people.

f. (Ideally) Identify a professional producer of video materials to take the raw footage and produce the final product.

II. Design a training workshop for community resource people that will provide a forum for sharing ideas, experiences, and expertise in collaboration with students, teachers, and schools.

a. Identify teachers, administrators, and students at target grade levels to assist in planning and presenting.

b. dentify members of the community who have successful experience in working with schools to assist in planning and presenting.

c. Discuss video-taped examples to develop presentation techniques and activities that utilize successful practices.

d. Discuss the variety of ways to form partnerships with students, teachers, and schools (outlined in handbook).

e. Locate times and places to hold sessions that would meet the needs of potential participants.

f. Design a marketing strategy for promoting the value of the workshops.

III. Design a similar training workshop for teachers.

a. Identify teachers who are experienced and successful in using outside resources to assist in planning and presenting.

b. Identify members of the community who have successful experience in working with schools to assist in planning and presenting.

c. Present methods for instructional planning that includes the use of community resources (outlined in handbook).

d. Identify ways teachers locate appropriate community resources and minimize red tape and time delays (outlined in handbook).

e. Discuss video-taped examples to identify ways to prepare students for speakers or site visits.

---

1993). Effects on the business partner can include maintenance and quality of involvement with schools and employee knowledge of and attitudes toward education. There is positive qualitative data of effects on teaching strategies (Farrell, 1992), networking (Beutel, Khashabi, & Marriott, 1991), curriculum development (Clark, 1990), and teacher self-esteem and job satisfaction (Ehrman, Treadwell, & Young, 1991; Gottfried, Brown, Markovits, & Changar, 1993). There is virtually no research or evalu-ation on student or district effects; however, new programs are beginning to include components to study these dimensions of the experience (Gottfried et al., 1993).

A compilation of ideas from several programs involving partnerships between schools and the technical business community (Heath, 1990; National Association of Partners in Education, 1990; North Carolina Museum of Life and Science, 1990; Pacific Northwest Laboratory, 1993-94) suggests an outline for developing effective partnerships. Table 1 describes a three-point program for improving interactions among community resource people, teachers, and students. These points include creating a resource of successful practices, implementing a training and idea-sharing program for community resource people, and implementing a similar program for teachers. The items under each of the three components should be used to identify observable characteristics of proposed programs that can form the basis for evaluation.

The approach is aimed at maximizing successful interactions between business and education partners when working directly with students. Direct contact with students is the most risky and complex type of connection with outside expertise but one that offers the greatest potential benefits. Teachers regularly express their concern that students will not behave in a way that results in the best use of a community resource. Conversely, teachers are concerned that an outside resource person will talk down to the students or over their heads and not involve them in constructive ways during visits and presentations. In many cases, these concerns inhibit teachers from investing the extra time needed to utilize outside resources (Flick et al., 1995).

Larger partnerships have begun to address some of these concerns by creating training programs and evaluation components that aim to maximize value and rewards of the interactions among students, teachers, and the outside experts. The Pacific Northwest Laboratory (1993–1994) operates a set of programs for sharing scientific expertise with schools that involves speaker training and a short-term evaluation component. The Science Advisors program of Sandia National Laboratories (Heath, 1990) is considering the development of video tapes of successful presentations for use in training new speakers. Our own experience with

Lawrence Livermore National Laboratories promises to develop programs that take a critical look at how well the scientific experts and teachers collaborate to enhance the benefits to students while meeting higher standards in science education.

## References

American Association for the Advancement of Science (AAAS), Project 2061 (1993). *Benchmarks for science literacy*. New York: Oxford University Press.

Atkin, J.M., & Atkin, A. (1989). *Improving science education through local alliances*. Santa Cruz, CA: Network Publications.

Beutel, C., Khashabi, D., & Marriott, S. (1991, October). Teacher voice project. *Proceedings of the National Conference of Scientific Work Experience Programs for K–12 Teachers*. Berkeley, CA: University of California.

Bureau of Education (1918). *Cardinal principles of secondary education*. Washington DC: Government Printing Office.

Clark, R.J. (1990). Extending the boundaries of teacher education through corporate internships. *Journal of Teacher Education, 41*(1), 71–76.

Ehrman, P., Treadwell, G. & Young, J. (1991, October). Maximizing our impact on science and mathematics education. *Proceedings of the National Conference of Scientific Work Experience Programs for K–12 Teachers*. Berkeley, CA: University of California.

Farrell, A.M. (1992, March). What teachers can learn from industry internships. *Educational Leadership, 49*(6), 38–39.

Flick, L., Fekete, D., Hawkins, B.H., & Stone, R.H. (in press). Teacher use of community resources in the development of business, industry, and education partnerships. *Science Educator*.

Gottfried, S.S., Brown, C.W., Markovits, P.S., & Changar, J.B. (1993). Scientific work experience programs for science teachers: A focus on research-related internships. *Excellence in edu-*

cating teachers of science: AETS Yearbook. Columbus, OH: ERIC/CSMEE (SE 053469).

Heath, R.B. (1990). *Science advisors (SCIAD) program quarterly report for Oct.–Dec. 1990*. Albuquerque, NM: Sandia National Laboratories.

Kubota, C. (1993, March). Education-business partnerships: Scientific work experience programs. *Digest*. Columbus, OH: ERIC/CSMEE. (EDOSE93-3).

Maeroff, G.I. (1988). *The empowerment of teachers: Overcoming the crisis of confidence*. New York: Teachers College Press.

National Association of Partners in Education (1990). *A practical guide to creating and managing a business/education partnership*. Alexandria, VA: Author.

National Commission on Excellence (1983). *A nation at risk*. Washington, D. C.: U.S. Government Printing Office.

National Educational Goals Panel (1991). *National educational goals report: Building a nation of learners*. Washington, D. C.: U.S. Government Printing Office.

National Research Council (1996). *National science education standards*. Washington, DC: National Academy Press.

North Carolina Museum of Life and Science. (1990). *Sharing science: Linking students with scientists and engineers*. Durham, NC: Author.

Northwest Regional Educational Laboratory (NWREL) (1994/95, Winter). *Catalyst: Quarterly newsletter of the Northwest Consortium for Mathematics and Science Teaching*. Portland, OR: Author.

Pacific Northwest Laboratory. (1993–1994). Guide to science, mathematics, engineering, and technology education programs at Pacific Northwest Laboratory. Richland, WA: Author.

U.S. Department of Labor. (1991). *What work requires of schools: A SCANS report for America 2000*. Washington, DC: U.S. Government Printing Office.

## Author Note

**Lawrence B. Flick** is an assistant professor in the Science and Mathematics Education Department at Oregon State University. His research interests include the nature of inquiry-oriented instruction and instructional applications of technology. He has served on various boards for the National Association for Research in Science Teaching and the Council for Elementary Science International.

**Elizabeth Lambert** is program manager for the Linn-Benton Business Education Compact, a partnership building organization. She works to connect the business community with local teachers. These partnerships range from math and science curriculum enhancements to teacher internships. She has also been a primary author of a state-wide work-based learning manual for businesses in Oregon.

**Norman G. Lederman** is an associate professor in the science and mathematics education department at Oregon State University. He has published over 50 research articles and is a consistent presenter at NSTA conventions. He is past-president of the Association for the Education of Teachers in Science and a former member of the NSTA Board of Directors.

# A Paradigm for Developing a Demonstration Classroom Program

Julie L. Wilson
Edward L. Pizzini

Professional development for science teachers can consist of national and local conventions, district workshops, or local college or university courses. This diversity results in science educators having a variety of professional development opportunities in which to participate. The number of opportunities provided as well as the number of educators who attend them indicates that these practices are both acknowledged and accepted. Educators acknowledge them as important to their professional development, and staff developers accept the format as significant to improving science education in the classroom. Yet alternative practices are still worth considering, specifically demonstration classrooms that are combined with long-term staff development. This article will briefly examine the development of such a program, and then specifically suggest areas that staff developers should consider when coordinating a demonstration classroom effort.

## DEVELOPING A DEMONSTRATION CLASSROOM PROGRAM

A demonstration classroom is a place where a teacher who is learning a new instructional strategy can observe a teacher who has expertise in that method. For example, a teacher who is interested in learning about problem solving would visit a problem-solving expert's classroom to observe the practice as well as to interact with the experienced teacher and students.

The notion of a demonstration classroom program originated from a survey conducted in Iowa. When educators were asked about their needs for professional development, they specifically expressed the desire to see one another in their classrooms (Sweeney, Kemis, Lively, & Sorenson, 1992). Although the request seemed new and novel, it was hardly either. University laboratory schools and professional development schools have specifically used demonstration classrooms as places for educators to see and experience instructional practices. While several educators could benefit from visiting these classrooms, historically only those involved in preservice training and professional development schools have attended. By examining the previous demonstration classroom model and modifying it to meet current educational needs, a new program emerged that made demonstration classrooms available to all educators.

Demonstration classroom programs became an advocated alternative professional development practice in Iowa for three years. Throughout the development and the enactment process, numerous conferences contemplated demonstration classroom use and presented successful models (e.g., Governor's II Conference on Math and Science, Iowa Science Teachers Conference, and The Iowa Academy of Science Meeting). Funds were allocated from the Iowa Math and Science Eisenhower Program, and the combined efforts of state, regional, and district science educators were common place. At the height of this effort over 20 different demonstration classrooms became available to educators throughout the state.

To support the demonstration classroom visit an inservice component was added. The goal of combining the two was to provide inservice participants with an active example of the methodology being advocated and to provide long-term support as they incorporated the strategy in their classrooms. Numerous conferences and meetings contemplated a variety of demonstration classroom and inservice formats. The organizers at one demonstration site began with an inservice that introduced the problem-solving methodology in the summer, proceeded to multiple visitations to a demonstration classroom and follow-up during the academic year, and concluded with a final follow-up session in early summer. Follow-up sessions consisted of meetings with the entire group at specified locations or individual visits to teachers in their classrooms. In either case, follow-up sessions addressed the immediate concerns and questions of participants in the demonstration program.

## A PARADIGM FOR DEVELOPING A DEMONSTRATION CLASSROOM PROGRAM

Universities, local school districts, and area education agencies (organizations that service multiple school districts in rural areas) all evolved in their use of demonstration classrooms. During both the development and the implementation, these orga-

nizers identified several considerations that are salient in the paradigm of developing demonstration classroom programs. These considerations have been formed into specific guidelines that are important to demonstration classroom program development.

## Reforms

Demonstration classrooms should reflect the reform initiatives that are being advocated at a state and national level. Current science reforms at the national level include *Benchmarks for Science Literacy* (AAAS, 1993) and the *National Science Education Standards* (NRC, 1996). Both *Benchmarks* and *Science Standards* provide curricular, instructional, and assessment guidelines for teachers. They encourage science as inquiry, demonstrate a connection to other disciplines, and apply science to the surrounding environment (AAAS, 1993; NRC, 1996). State initiatives are often based on these national reforms and are crafted to reflect them (e.g., Arizona Science Essential Skills, 1990). At any level this means teaching with respect to how students learn and providing opportunities to actively engage all students in the learning environment. Thus, demonstration classrooms should be active learning environments that utilize students' current science knowledge while incorporating other disciplines.

Reforms in a demonstration classroom can also provide a platform for instructional enhancement for both the demonstration teacher and the visiting educator. Demonstration teachers must reflect critically on their practices and share these insights with others, while visiting educators observe the interpretations, analyze them, and construct their own meanings. As both share observations and thoughts, they each gain a greater understanding of the science reforms and the implications for instruction in science classroom.

## Focus

Demonstration classrooms should have a clearly defined focus. As educators know, clarity is important for effective teaching (Acheson & Gall, 1992; Porter & Brophy, 1988). Demonstration classrooms will need to be clear on their instructional objectives if they are to be effective. Anything could be selected as a demonstration classroom focus: instructional strategies, curricular enhancements, classroom management techniques, questioning strategies, alternative assessment procedures, or themes.

In Iowa, several demonstration classrooms promoted a problem solving technique. Educators who visited these classrooms knew they would see the methodology and the supporting instructional practices. If they had previously tried the practice, they would have an opportunity to compare and refine their instruction. In addition, teachers who attended the problem solving demonstration classrooms created informal connections with other educators who held the similar instructional interests. A focus should not be limiting, but it should provide a specific direction that allows science educators to address their instructional needs.

## Needs

Demonstration classrooms should meet specific needs of the people they seek to service. Fullan with Stiegelbauer (1991) found the greatest failures of inservice were due to (a) the topics being selected by people other than those for whom the inservice is intended, (b) inservices rarely addressing individual needs and concerns, and (c) involvement of teachers from many schools and/or districts. Demonstration classrooms are no exception to this. To be truly effective demonstration classrooms should be developed in response to the needs of the associated school and/or district. An example of this can be found in Iowa City, Iowa. In the Iowa City Community School District there has always been an interest in

math and science integration. Problem solving methodologies have been viewed as a vehicle to promote this connection. Local interests led to the development of a demonstration site that promoted math/science integration through problem solving. Due to the expressed interest of teachers driving the development of this site, these classrooms were used extensively.

## Clinical Supervision

Demonstration classrooms should utilize the clinical supervision model, which includes a pre-conference, an observation, and a post-conference. Acheson and Gall (1992) have provided the framework that was used at several Iowa demonstration classroom sites. During the pre-conference, demonstration teachers discussed the upcoming demonstration, reflected on previous classroom lessons, and provided visiting educators with a focus during the observation. This information provided clarity and specificity for the observer. During the observation, the visiting educator collected data in either a qualitative or a quantitative manner. Observers counted teacher/student behaviors or recorded verbatim exchanges. The focus was formative in style, and the emphasis was on in-depth examination. The post-conference followed the observation and was a time for reflection about the observation, discussion of data collected, and planning for the initiation of the formulated strategies. Demonstration classrooms that utilized the clinical supervision model had teachers who reflected on their practice, addressed their personal needs, and eased their fears in implementing a new strategy (Wilson, 1994).

## Demonstration Classroom Program

Demonstration classrooms should coordinate with inservice. If the goal is to have profound and enduring change in the science classroom, one time visits will not be successful. Research has concluded that to be most effective, inservice training should include theory, demonstration, practice, feedback, and classroom application (Joyce & Showers, 1988). An inservice with these would incorporate the components of long-term support, peer coaching, team building, collegial support, reflection, clinical supervision, and modeling (Acheson & Gall, 1992; Fullan & Stiegelbauer, 1991; Joyce & Showers, 1995; O'Brien, 1992; Showers, 1985). This theoretical framework can be facilitated through a long-term inservice that incorporates a classroom that demonstrates the methodology, provides opportunities for ongoing visits, utilizes pre- and post-conferences with the visitation, and promotes interactions among all professionals. This type of professional development structure becomes a demonstration classroom inservice. By incorporating a demonstration classroom into an inservice program, it is possible to promote profound and enduring change in science education.

## Collaboration

Demonstration classroom programs need collaboration between universities, area educational agencies, and local school districts during development and implementation.

A review of the literature suggests that there are two reasons for encouraging this collaboration: (a) vertical teams allow for a depth of support for the innovation, and (b) by involving all members, each contributes expertise that supports the change. Fullan and Stiegelbauer (1991) and Joyce and Showers (1995) continually reinforce the idea that all levels need to be committed to the change. Pink and Hyde (1992) in a review of case studies found that having universities, districts, and teachers collaborate was important to successful change. They found that the university personnel were skilled outside change agents who fostered a collaborative and supportive environment for the teachers and administrators.

In Iowa, district personnel were able to make financial contributions, arrange the scheduling, provide materials, offer personal support, and furnish release time for all participants. The ultimate message from the district to the participating teacher was one of support. Teachers became sources of information about what did and did not work. They furnished ideas and enthusiasm that offered a new direction to the demonstration classroom program.

Collaboration allows each member to contribute an important perspective and be part of a team committed to the improvement of science education. It also allows depth and breadth of commitment. In Iowa, each member was critical to the development of demonstration classrooms; without any one component the classrooms would not exist as educators knew them.

## Time

Demonstration classroom programs require a substantial amount of time for development, ample time for all people involved, and adequate time for various program components. In Iowa, most demonstration classroom organizers and participants would agree that there was never enough time. Several demonstration classroom sites were implemented after 10 months of informational meetings, workshops, and team work sessions. Each meeting focused on new problems and issues, and each meeting addressed the evolution of the demonstration classroom. Teachers specifically needed ample time to participate in all parameters of the demonstration classroom. Personnel from the school administration, district office, area education agency office, and university also found themselves with unique tasks that required coordination with others. Another time factor that needs consideration is the duration and scheduling of the demonstration classroom: When will demonstration classrooms occur? Will there be a series of demonstrations or will they be spaced out over the year? How long should the visitation blocks be? Further considerations include the time participating teachers will spend traveling to inservice activities, demonstration visits, and follow-up activities. In Iowa, the investment of time has paralleled the commitment to the development of specific demonstration classroom programs.

## Change

Demonstration classroom programs should emphasize and support change. Critical to the first steps of change are encounters that create conflict between actual and espoused theories, cause perturbations, or foster a new awareness (Etchberger & Shaw, 1992; Hall, Wallace, & Dossett, 1973; Sergiovanni & Starratt, 1993). In resolving or addressing these, there is active engagement in change. Loucks and Pratt (1979), in their lasting wisdom, state the four assumptions that need to be acknowledged during a period of change: (a) change is a process, (b) change is accomplished by individuals, not institutions, (c) change is a highly personal experience, and (d) change entails developmental growth in both feelings and skills in using new programs.

Participants, district personnel, staff developers, and university counterparts are all subject to change during the demonstration classroom effort. Successfully engaging in this process requires a supportive environment and informative resources. Creating a supportive climate entails encouraging broad participation—from teacher to district administrator to university professor. As each person is involved they should also have access to information about the process of change. This should be available during workshops and conferences and specifically address the change process and the time change takes.

## Evaluation

Demonstration classroom programs should engage in ongoing evaluation. This is essential for both teacher practice and inservice structure. Practitioners need to have opportunities to evaluate their current practice, reflect on the changes needed, and modify accordingly. (Sergiovanni & Starratt, 1993; Schön, 1983; Baird et al, 1991.) Demonstration classroom programs, as previously suggested, can provide educators with this opportunity. Inservice programs should evaluate their successes and failures, reflect upon the changes needed, and modify their format. Yet most inservice programs have limited evaluation efforts; assessment programs are brief in duration, assessment instruments are often limited in the what they measure, assessment techniques are often pre- and post-measures, and assessment is rarely used for the modification of a program. To learn more about the potential of demonstration classroom programs and to allow staff organizers to adjust their programs, assessment should be embedded throughout all efforts. Evaluation, reflection, and modification are critical to inservice coordinators and educators in creating the most beneficial inservice.

## DEMONSTRATION CLASSROOM PROGRAMS

Demonstration classroom programs combine inservice with visits to a classroom that is actively participating in the advocated methodology. As educators visit demonstration classrooms they experience the successes and frustrations of both the students and the demonstration teacher. During this process participants have multiple opportunities to address their personal needs about the enacted practice, they engage in conversations about practice with other professionals, and they develop a sense of the practice as a strategy in the classroom (Wilson, 1994). The demonstration classroom adds a new dimension to inservice practice. Specifically, "Staff development supports the demonstration and the demonstration clarifies the inservice" (Wilson & Pizzini, 1994, p. 7).

National Science Teachers Association

## CONCLUSION

The paradigm presented here is not a prescription for demonstration classroom developers; instead it is a description about a possible path to take. Staff developers, administrators, and educators can take each recommendation and craft it to meet the needs of their site. This process hinges on being a collaborative effort that puts the science teacher at the center of every decision. With reflective and thoughtful planning, distinct features of effective staff development can thread throughout the demonstration program. With commitment and support, this program can reduce the isolation between educational professionals and a create sense of community among the participating group of educators. The beneficiaries of a carefully planned and executed demonstration classroom paradigm will be the teachers and, ultimately, the students.

## References

Acheson, K.A., & Gall, M.D. (1992). *Techniques in the clinical supervision of teachers: Preservice and inservice applications.* New York, NY: Longman.

American Association for the Advancement of Science (AAAS) (1993). *Benchmarks for science literacy.* New York, NY: Oxford University Press.

Arizona Department of Education (1990). *Arizona Science Essential Skills.* Phoenix, AZ: Author.

Baird, J.R., Fensham, P.J., Gunston, R.F., & White, R.T. (1991). The importance of reflection in improving science teaching and learning. *Journal of Research in Science Teaching, 28*(2). 163–182.

Etchberger, M.L., & Shaw, K.L. (1992). Teacher change as a progression of transitional images. A chronology of a developing constructivist teacher. *School Science and Mathematics, 92*(8), 411–417.

Fullan, M. G. (with Stiegelbauer, S.) (1991) *The new meaning of educational change.* New York, NY: Teachers College Press.

Hall, G.E., Wallace, R.C., & Dossett, W.A. (1973). *A developmental conceptualization of the adoption process within educational institutions.* Austin, TX: Research and Development Center for Teacher Education, University of Texas.

Joyce, B.R., & Showers, B. (1995). *Student achievement through staff development:* Fundamentals of school renewal. New York, NY: Longman.

Loucks, S.L. & Pratt, H. (1979). A concerns-based approach to curriculum change. *Educational Leadership, 34*(4), 212–215.

National Research Council (NRC) (1996). *National Science Education Standards.* Washington, D.C.: National Academy Press.

O'Brien, T. (1992). Science inservice workshops that work for elementary teachers. *School Science and Mathematics, 92*(8), 422–426.

Pink, W.T., & Hyde, A.A. (Eds.) (1992). *Effective staff development for school change.* Norwood, NJ: Ablex.

Porter, A.C., & Brophy, J. (1988). Synthesis of research on good teaching: Insights from the work of the Institute for Research and Teaching. *Educational Leadership, 45*(8), 74–85.

Schön, D.A. (1983). *The reflective practitioner.* New York, NY: Basic Books.

Sergiovanni, T.J., & Starratt, R.J. (1993). *Supervision: A redefinition.* New York, NY: McGraw-Hill, Inc.

Showers, B. (1985). Teachers coaching teachers. *Educational Leadership, 2*(7), 43–48.

Sweeney, J., Kemis, M., Lively, M., & Sorenson, C. (1992). *A summary of the assessment of elementary and secondary curriculum needs and supply and demand for teachers in mathematics and science.* Ames, IA: Research Institute for Studies in Education.

Wilson, J.L.(1994). *The effects of demonstration classroom on elementary science teachers involved in a problem solving inservice program.* Unpublished doctoral dissertation, University of Iowa, Iowa City.

Wilson, J.L. & Pizzini, E.L. (1994). A new perspective for science inservice: Problem Solving Demonstration Classrooms. *Iowa Science Teachers Journal, 30*(3), 3–11.

## Author Note

**Edward L. Pizzini** is a professor of science education at the University of Iowa. He is the developer of the Search, Solve, Create, and Share (SSCS) Model of Problem Solving, which in 1994 was recognized as a National Program of Excellence. He is a two-time winner of NSTA's OHAUS award.

**Julie L. Wilson** is an assistant professor of science education at the University of Arizona. She worked extensively on the development and assessment of the Problem Solving Demonstration Classroom program in Iowa. Currently she is directing an Eisenhower science teacher enhancement program and is focusing on multicultural science education.

# Past, Present, and Future: Mathematics and Science Education Through a Statewide Network

Gerry M. Madrazo, Jr.
Gretchen Van Vliet

As the 1984 report of the North Carolina Commission on Education for Economic Growth quoted North Carolina Governor Charles B. Aycock,

Everyone must recognize that the wealth of the State is dependent upon the wages which are paid to the earners, and these wages in turn are dependent upon the capacity of the wage earners. And this capacity is dependent in a large measure upon the quickness and skill which come from an acquaintance with books. (NC Commission on Education for Economic Growth, 1984, p. 1)

Aycock linked education with economic growth in 1901 as did the Commission in 1984, 80 years later when meeting to discuss improvements in education in North Carolina. The Commission reported a "new industrial revolution" (p. 2) in North Carolina and said future efforts to attract new businesses to the state could not "depend simply on a plentiful supply of strong backs and willing hands. If we are to cope successfully with economic change," reported the Commission, "our state must be able to offer potential employers a labor force that is not only strong and eager, but also well educated and highly adaptable." (p. 3)

Due to this attitude in 1984, a mathematics and science education project that had originated in 1981 at the University of North Carolina-Chapel Hill found state support and became the University of North Carolina Mathematics and Science Education Network, a network that would cover the state of North Carolina, further educating math and science teachers and reaching under-represented groups of students in the maths and sciences. At the impetus of UNC-Chapel Hill Chancellor Christopher Fordham, several professors in the School of Education were charged with developing a program that would appropriately certify math and science teachers in the state in order to bring them up to the level of performance needed to teach the classes they were already teaching. During 1980-81 concern had arisen due to a lack of qualified math and science teachers in the state, with universities certifying only 167 mathematics teachers for 620 mathematics teaching positions and only 218 science teachers for 310 science positions. Less than one half of the math teachers in the middle and high schools were properly certified to teach one or more math classes, while less than 40 percent of middle and high school science teachers were appropriately qualified to teach science. (Stedman, 1983, p. 3)

The original focus of the math and science program was to be at the middle school level "since it's at the middle school level that the decision is often made by students to either continue with mathematics and science or stay as far away from it as they possibly can," explained Dr. Hunter Ballew, professor of Education at UNC-Chapel Hill and key network planner, in an interview conducted a year after the official forming of the network. (The Network, 1985, p. 2) In 1981, Chancellor Fordham used discretionary funds from his office to establish a local mathematics and science education center that offered summer workshops for middle school teachers. Teachers participated in a two-week summer program and then spent one day each week throughout the school year at the university. Dr. Fordham, in an interview a year after the forming of the network, discussed how the network arose from a summer program for middle school mathematics and science teachers. "It seemed to be such a success that we continued it subsequently, and it caught on in terms of the idea that it ought to be expanded and reach across the state and that's how the network came into being." (The Network, 1985, p. 3)

In December 1982, Chancellor Fordham submitted a proposal for four centers that would offer two curricula: a 16-semester hour curriculum for middle school teachers and a 36-hour curriculum for high school teachers. While this was happening in Chapel Hill, Chancellor E. K. Fretwell and Dean Sherman Burson of Arts and Sciences at UNC-Charlotte were part of an interdisciplinary faculty committee planning out science and mathematics education initiatives. This committee and Discovery Place Museum held a program of workshops for teachers.

The Committee on Science and Mathematics Education of the North Carolina Board of Science and Technology and the State Board of Education submitted an initial proposal to formally establish a "Network of Mathematics and Science Education Centers" to Chancellor Fordham and Dr. William Friday, President of the University of North Carolina in May 1983. Within this atmosphere of educational need, the UNC General Administration established the University of North Carolina Mathematics and Science Education Network, to provide continuing education for public school teachers with the direct aim of increasing the pool of math and science teachers in North Carolina.

The initial goals of the network outlined by Stedman (1983) were to

1. Increase the qualifications of those mathematics and science teachers already teaching in North Carolina public schools.

2. "Develop and evaluate" professional development programs that could be replicated at universities in the state.

3. Sponsor basic research and evaluation in mathematics and science education.

4. Increase the effective use of educational technologies in all schools. (p. 3)

A key endorsement in January 1984 from Governor Hunt added an "important source of support for the network's development." (Stedman, 1984, p. 1) In early summer 1984 the North Carolina General Assembly appropriated permanent funding to expand and improve the network. On July 12, 1984, the UNC Board of Governors and General Administration formally established the Mathematics and Science Education Network as it exists today, with eight teacher education centers on state university campuses across North Carolina, a research and development center at North Carolina State University, and a liaison center at the North Carolina School of Science and Mathematics. The eight centers are located at Appalachian State University, East Carolina

University, Fayetteville State University, North Carolina A&T State University/ UNC-Greensboro, UNC-Chapel Hill, UNC-Charlotte, UNC-Wilmington, and Western Carolina University.

As a result of several years of planning, the network was established to "significantly upgrade" mathematics and science education in North Carolina public schools. (The Network, 1984, p. 3) The network was seen as the most effective way to meet two fundamental educational needs. The first is the necessity to develop close working relationships among educators, industrial scientists, university faculty, and business leaders. The centers provide the means to mobilize these local partnerships, which can provide key resources to public education. The second need is to effectively disseminate successful educational improvements, such as innovative educational technologies, new instructional materials, model development programs, or other activities. (p. 3)

## THE PRE-COLLEGE PROGRAM

With the network up and running, the first programs offered through the centers in 1984-85 educated almost 2,800 teachers through professional development activities. The next year saw the start of the second major branch of the network: the Pre-College Program. The program, funded originally by the Carnegie Foundation and the National Action Council for Minorities in Engineering, was designed to increase the number of historically underrepresented students—minorities and females—pursuing fields in mathematics and science at the university level and subsequently moving into careers in mathematics, science, technology, engineering, and teaching. The program is located at six universities, five of which are center sites: Elizabeth City State University, Fayetteville State University, NC A&T State University, North Carolina State University, UNC-Chapel Hill, and UNC-Charlotte.

The Pre-College Program provides students in grades 6–12 with academic enrichment activities and works with teachers to increase their knowledge of mathematics and science and their understanding of the needs of underrepresented students. The six components of the program include (a) academic enrichment classes, (b) Saturday Academy sessions, (c) Summer Scholars program, (d) Parents Involved in Education (PIE) groups, (e) teacher in-service workshops, and (f) leadership and career awareness activities. (The Network, 1995, p. 5–6)

The academic enrichment classes involve daily classes with science labs, field trips for experiential learning, individualized tutoring, and career counseling. The Saturday sessions during the school year allow students to rotate through classes in mathematics, science, communication skills, and self-esteem/career awareness. The instructors are from university faculties, public schools, and businesses and industries. The Summer Scholars program is held on each of the six university campuses and meets for 100 class hours during the summer; a greater variety of classes are offered than during the Saturday Academy sessions. This component has received National Science Foundation Young Scholars funding to support university faculty and enrichment activities for the middle school portion of this session. The PIE program involves the parents of the Pre-College students. Parents help raise funds for activities and assure that their children are fully participating in the program activities. Pre-College teachers go through in-service workshops to learn methods and materials that encourage minorities and females to pursue mathematics and science and teach the teachers to establish bias-free classrooms. The last component, leadership and career awareness activities, includes career counseling, role-model speakers, and field trips.

Each spring a Pre-College Day is held, allowing students from the different Pre-College sites to compete for awards in mathematics, science, a quiz bowl, writing, public speaking, and a poster and design competition. Students also hear speakers on everything from dentistry to medicine to social work. The Pre-College Day is hosted by a different site each year, and the host site involves parent, student, and staff volunteers acting as exam proctors, test score recorders, and guides on the university campus.

By the end of the 1993–94 school year, over 3,000 students from across the state were participating in the Pre-College Program, representing 26 school districts and 77 schools. A follow-up survey done in the fall of 1994 of Pre-College graduates from 1991, 1992, 1993, and 1994 showed that of the 82 percent of the graduates who were contacted, 98 percent were enrolled in college and 60 percent were majoring in a mathematics- or science-related field.

## CENTER ACTIVITIES

Each center carries out various mathematics and science professional development workshops for teachers, offering programs that either are fulfilling specific needs of their particular area in the state or are funded by grants written by center staff members. Due to the locations of the centers, teachers in each area of the state can be directly served with professional development opportunities.

For the tenth year the network and the State Department of Public Instruction (SDPI) offered summer institutes to K–8 mathematics and science teachers. The institutes are offered through the centers from June through December, ranging from *Problem Solving Data Analysis* at the center at East Carolina University to *Patterns, Cycles, & Change: Science in the Middle School Curriculum* at the center at UNC-Chapel Hill to *Integrating Life Science in Middle Grades Science* at the center at Appalachian State University. Teachers receive a $300 stipend for workshop participation and either graduate

credit or recertification credit. The characteristics of the summer institutes are (a) an emphasis on integration of mathematics and science and the development of thinking, reasoning, and problem-solving skills, (b) an overview of the SDPI Standard Course of Study in mathematics and/or science, (c) a team-teaching approach, whereby both a university faculty member and a master teacher from a local school jointly instruct each course, (d) laboratory and field work, and (e) a partnership among various components of the university, local education agencies, the Department of Public Instruction, and local organizations such as museums and businesses (The Network, January 1995, p. 2). During the 1995 summer institutes, 480 teachers enrolled. The center at UNC-Charlotte conducted an institute entitled *Mesozoic North Carolina: Geology in the Time of the Dinosaurs*, a "field-intensive, hands-on, experiential" course on the principles of geology using the Mesozoic Era to study the geology of North Carolina. The course consisted almost entirely of time spent in field, traveling around North Carolina and studying the types of geology found in the state.

In 1994, UNC-Wilmington's Mathematics and Science Education Center instituted the Technology Loan Program (TLP) in cooperation with the Southeast Partnership of the North Carolina Science and Mathematics Alliance. The TLP consists of a series of training workshops in which teachers learn to use biotechnology kits, Personal Science Laboratory kits, water quality test kits, and Star Lab for astronomy. Teachers use the kits to teach students computerized data acquisition, recombinant DNA technology, water quality studies, video microscopy, and astronomy. Once teachers complete the workshops they are qualified to participate in the loan of the TLP kits for up to two weeks for use in their classrooms.

The VISION program, Vision of Industry and Schools Initiating an Ongoing Network, sponsored by the Semiconductor Research Corporation Competitiveness Foundation and the Charlotte Region Partnership of the North Carolina Science and Mathematics Alliance and the Centers at North Carolina State University and UNC-Charlotte, is based on partnerships among industry, university, and local schools. The VISION project provides science and mathematics teachers with the opportunity to experience firsthand the industrial uses of the subjects they teach every day. Participating companies include Carolina's Medical Center, Duke Power Company, Hankel Corporation, Hoechst-Celanese, Westinghouse Corporation, Micro Computing of North Carolina (MCNC), Research Triangle Institute, E. I. Dupont de Nemours & Company, Inc., IBM, Northern Telecom, Baxter Healthcare Corporation, and The Virkler Corporation. Teachers who have been involved in this program have explored the process of computer chip making, software design for military and commercial aircraft, the manufacturing of the IBM PS2, and networking and communication systems. As a result of the program at the center at UNC-Charlotte, a booklet of classroom lessons was developed to enhance the teaching of mathematics and science through industry applications. The contributors to the curriculum included teachers from four school systems. Such lessons as *Color Exploration*, *Recycling Chemicals to Preserve the Environment*, *Physical Properties of Polyester Fiber*, *Atomic Spectra*, *The Making of Soap*, and *Graphing in Physical Science and Industry* resulted from the experiences of the teachers.

## A VISION OF SYSTEMIC REFORM

Several organizations are involved in efforts to improve science and mathematics education. The UNC Network is one such institution. It is, until now, the only university outreach program of its kind dedicated to improving K–12 science and mathematics teaching and learning in the state. With the national effort toward systemic reform, the network is in a strategic position to pro-

vide leadership and to collaborate with other institutions and partners toward this end. The "bottom line" is improvement of student achievement in science, mathematics, and related fields. It is imperative that collaboration among all stakeholders must occur. They must include teachers, educational leaders, scientists, business and community leaders, policymakers, parents, and students. When these efforts are guided by mathematics and science education standards, the reform tasks can move productively in the same general direction.

The National Science Foundation (NSF) has provided certain criteria representing significant implementation activities for statewide systemic reform initiatives. (NSF, 1993; 1995) They are comprehensive, but include some of the following elements:

1. Curriculum, assessment, and professional development—use of national frameworks, standards, or benchmarks; identification of curriculum materials and instructional strategies; use of data that will be used to assess improvements in student learning.

2. Enabling policy—alignment of the systemic program with federal, state, and local reform efforts; policies that support long-term, continuous professional development for teachers.

3. Issue of equity—students, regardless of gender, ethnicity, disability, and poverty levels are served by the program; teacher participants represent diverse backgrounds; clear understanding of "performance gaps" and plans to ensure that all students have an equal opportunity to learn.

4. Establishing partnerships—involvement of all stakeholders with the program.

5. Articulation between K–12 and higher education.

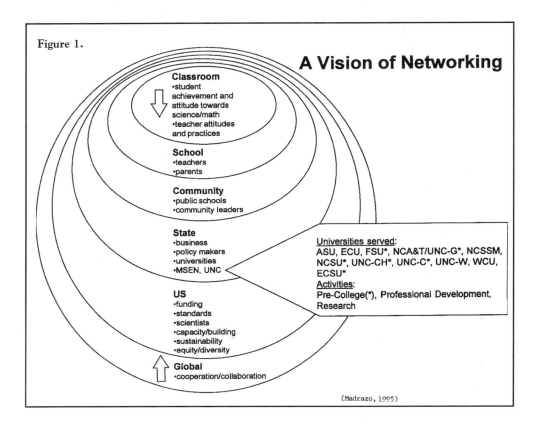

**Figure 1.**

## A Vision of Networking

**Classroom**
- student achievement and attitude towards science/math
- teacher attitudes and practices

**School**
- teachers
- parents

**Community**
- public schools
- community leaders

**State**
- business
- policy makers
- universities
- MSEN, UNC

**US**
- funding
- standards
- scientists
- capacity/building
- sustainability
- equity/diversity

**Global**
- cooperation/collaboration

Universities served:
ASU, ECU, FSU*, NCA&T/UNC-G*, NCSSM, NCSU*, UNC-CH*, UNC-C*, UNC-W, WCU, ECSU*
Activities:
Pre-College(*), Professional Development, Research

(Madrazo, 1995)

6. Management and governance—including well-managed and cost-effective funding and strategies institutionalizing the effective components of the program.

7. Evaluation—an evaluation design that addresses the whole initiative and a system to monitor student performance.

## A CULTURE OF NETWORKING

When the network defines a construct of systemic networking among its university centers, it contributes a strategy to strengthen the infrastructure for science and mathematics education through alignment of resources and policies within the state (Figure 1). The commitment to change calls for a new image of mathematics and science education, which necessarily will include the integration of technology into both fields. Ultimately, the purpose is to move science and mathematics

into the framework of society and human affairs. To put it in the words of Francis Bacon: "Science reaches its greatest peak when it reaches the people...the ideal of human service is the ultimate goal of scientific effort—providing a better and more perfect use of human reason."

## References

D. J. Stedman (personal communication, January 31, 1984).

National Science Foundation. (1995). *Criteria for SSI midpoint review*. Arlington, VA: NSF.

National Science Foundation. (1993). *SSI; State-wide systemic initiatives in science, mathematics and engineering, 1993–94*. Arlington, VA: NSF.

North Carolina Commission on Education for Economic Growth. (1984, April). *Education for economic growth: An Action plan for North Caro-*

National Science Teachers Association

lina. (Available from the Office of the Governor, State Capitol, Raleigh, NC 27601).

Stedman, Donald J. (1983, May). *A proposal to establish a network of mathematics and science education centers in North Carolina.* Working paper presented to UNC President William Friday and NC Governor James Hunt.

UNC Mathematics and Science Education Network. (1984, December). *A network of centers to upgrade precollege science and mathematics education in North Carolina.* Discussion Paper.

UNC Mathematics and Science Education Network. (1985). Interview with Dr. Christopher Fordham, Chancellor at UNC-Chapel Hill.

UNC Mathematics and Science Education Network. (1985). Interview with Dr. Hunter Ballew, Professor of Education at UNC-Chapel Hill.

UNC Mathematics and Science Education Network. (1995, January). *Pre-college program status report.* (Available from the UNC Mathematics and Science Education Network, 134 1/2 E. Franklin Street, CB #3345, Chapel Hill, NC 27599–3345).

UNC Mathematics and Science Education Network. (October, 1994). *Summer Institutes 1995: Courses for mathematics and science teachers of grades K–8.* (Available from the UNC Mathematics and Science Education Network).

## Author Note

**Gerry M. Madrazo, Jr.** is the executive director of the University of North Carolina Mathematics and Science Education Network (MSEN) and clinical associate professor in the School of Education at the University of North Carolina at Chapel Hill. He has served as 1993 president of NSTA, president of the NC Science Education Leadership Association (formerly, NCSTA) as well as president of the North Carolina Science Teachers Association (NCSTA).

**Gretchen Van Vliet** is the information and communication specialist for the UNC Mathematics and Science Education Network (MSEN). She is a graduate student in the School of Journalism at the University of North Carolina at Chapel Hill and editor of the MSEN newsletter, the Golden Mean.